Communications
in Computer and Information Science   **1061**

Commenced Publication in 2007
Founding and Former Series Editors:
Phoebe Chen, Alfredo Cuzzocrea, Xiaoyong Du, Orhun Kara, Ting Liu,
Krishna M. Sivalingam, Dominik Ślęzak, Takashi Washio, and Xiaokang Yang

Editorial Board Members

More information about this series at http://www.springer.com/series/7899

Lemonia Ragia · Cédric Grueau ·
Robert Laurini (Eds.)

Geographical Information Systems Theory, Applications and Management

4th International Conference, GISTAM 2018
Funchal, Madeira, Portugal, March 17–19, 2018
Revised Selected Papers

 Springer

Editors
Lemonia Ragia
Technical University of Crete
Chania, Crete, Greece

Cédric Grueau
Polytechnic Institute of Setúbal
Setúbal, Portugal

Robert Laurini
Knowledge Systems Institute
Skokie, IL, USA

ISSN 1865-0929   ISSN 1865-0937 (electronic)
Communications in Computer and Information Science
ISBN 978-3-030-29947-7   ISBN 978-3-030-29948-4 (eBook)
https://doi.org/10.1007/978-3-030-29948-4

This Springer imprint is published by the registered company Springer Nature Switzerland AG
The registered company address is: Gewerbestrasse 11, 6330 Cham, Switzerland

Preface

The present book includes extended and revised versions of a set of selected papers from the 4th International Conference on Geographical Information Systems Theory, Applications and Management (GISTAM 2018), held in Funchal, Madeira, Portugal, during March 17–19.

GISTAM 2018 received 58 paper submissions from 31 countries, of which 12% were included in this book. The papers were selected by the event chairs and their selection was based on a number of criteria that included the classifications and comments provided by the Program Committee members, the session chairs' assessment, and also the program chairs' global view of all papers included in the technical program. The authors of selected papers were then invited to submit a revised and extended version of their papers having at least 30% innovative material.

GISTAM aims at creating a meeting point of researchers and practitioners to address new challenges in geo-spatial data sensing, observation, representation, processing, visualization, sharing, and managing, in all aspects concerning both information communication and technologies (ICT) as well as management information systems and knowledge-based systems. The conference welcomes original papers of either practical or theoretical nature, presenting research or applications, of specialized or interdisciplinary nature, addressing any aspect of geographic information systems and technologies.

The papers selected to be included in this book contribute to the understanding of relevant trends of current research on geographical information systems theory, applications and management, including: urban and regional planning, water information systems, geospatial information and technologies, spatio-temporal database management, decision support systems, energy information systems, and GPS and location detection.

We would like to thank all the authors for their contributions and also the reviewers who helped ensure the quality of this publication.

March 2018

Lemonia Ragia
Cédric Grueau
Robert Laurini

Organization

Conference Chair

Lemonia Ragia   Technical University of Crete, Greece

Program Chairs

Cédric Grueau   Polytechnic Institute of Setúbal - IPS, Portugal
Robert Laurini (Honorary)   Knowledge Systems Institute, France

Program Committee

Thierry Badard   Laval University, Canada
Jan Blachowski   Wroclaw University of Science and Technology,
                                 Poland
Thomas Blaschke   University of Salzburg, Austria
Alexander Brenning   Friedrich Schiller University, Germany
Jocelyn Chanussot   Grenoble Institute of Technology Institut
                                 Polytechnique de Grenoble, France
Filiberto Chiabrando   Politecnico di Torino - DAD, Italy
Keith Clarke   University of California, Santa Barbara, USA
Antonio Corral   University of Almeria, Spain
Joep Crompvoets   KU Leuven, Belgium
Paolo Dabove   Politecnico di Torino, Italy
Luís de Sousa   ISRIC - World Soil Information, The Netherlands
Anastasios Doulamis   National Technical University of Athens, Greece
Suzana Dragicevic   Simon Fraser University, Canada
Qingyun Du   Wuhan University, China
Arianna D'Ulizia   IRPPS, CNR, Italy
Ana Falcão   Instituto Superior Técnico, Portugal
Ana Fonseca   Laboratório Nacional de Engenharia Civil (LNEC),
                                 Portugal
Cidália Fonte   Coimbra University, INESC Coimbra, Portugal
Jinzhu Gao   University of the Pacific, USA
Lianru Gao   Chinese Academy of Sciences, China
Georg Gartner   Vienna University of Technology, Austria
Fabio Giulio Tonolo   ITHACA c/o SiTI, Italy
Luis Gomez-Chova   Universitat de València, Spain
Gil Gonçalves   University of Coimbra, INESC Coimbra, Portugal
José Gonçalves   Oporto University, Portugal
Cédric Grueau   Polytechnic Institute of Setúbal - IPS, Portugal
Hans Guesgen   Massey University, New Zealand

Bob Haining   University of Cambridge, UK
Stephen Hirtle   University of Pittsburgh, USA
Wen-Chen Hu   University of North Dakota, USA
Haosheng Huang   University of Zurich, Switzerland
Simon Jirka   52 North, Germany
Wolfgang Kainz   University of Vienna, Austria
Harry Kambezidis   National Observatory of Athens, Greece
Eric Kergosien   University Lille 3, France
Andreas Koch   University of Salzburg, Austria
Jacek Kozak   Jagiellonian University, Poland
Artur Krawczyk   AGH University of Science and Technology, Poland
Roberto Lattuada   myHealthbox, Italy
Robert Laurini   Knowledge Systems Institute, France
Jun Li   Sun Yat-Sen University, China
Songnian Li   Ryerson University, Canada
Christophe Lienert   ICA International Cartographic Association,
                                 Commission EWCM, Switzerland
Vladimir Lukin   Kharkov Aviation Institute, Ukraine
Paulo Marques   ISEL - ADEETC, Portugal
Gavin McArdle   University College Dublin, Ireland
Fernando Nardi   Università per Stranieri di Perugia, Italy
Anand Nayyar   Duy Tan University, Vietnam
Jose Pablo Suárez   Universidad de Las Palmas de Gran Canaria, Spain
Volker Paelke   Bremen University of Applied Sciences, Germany
Dimos Pantazis   University of West Attica, Greece
Massimiliano Pepe   Università degli Studi di Napoli Parthenope, Italy
Marco Piras   Politecnico di Torino, Italy
Alenka Poplin   Iowa State University, USA
Dimitris Potoglou   Cardiff University, UK
Lemonia Ragia   Technical University of Crete, Greece
Jorge Rocha   University of Minho, Portugal
Mathieu Roche   Cirad, France
Markus Schneider   University of Florida, USA
Sylvie Servigne   INSA Lyon, France
Yosio Shimabukuro   Instituto Nacional de Pesquisas Espaciais, Brazil
Francesco Soldovieri   Consiglio Nazionale delle Ricerche, Italy
Uwe Stilla   Technische Universität München, Germany
Jantien Stoter   Delft University of Technology, The Netherlands
Nicholas Tate   University of Leicester, UK
José Tenedório   Universidade NOVA de Lisboa, NOVA FCSH,
                                 Portugal
Ana Teodoro   Oporto University, Portugal
Theodoros Tzouramanis   University of the Aegean, Greece
Michael Vassilakopoulos   University of Thessaly, Greece
Lei Wang   Louisiana State University, USA
May Yuan   University of Texas at Dallas, USA

Additional Reviewers

Ludwig Hoegner	Technical University Munich, Germany
Sven Kralisch	University of Jena, Germany
Yusheng Xu	TUM, Germany

Invited Speakers

Uwe Stilla	Technische Universität München, Germany
Kostas E. Katsampalos	Aristotle University of Thessaloniki, Greece
Eugene Fiume	Simon Fraser University and University of Toronto, Canada

Contents

Reconstruction of Piecewise-Explicit Surfaces from Three-Dimensional Polylines and Heightmap Fragments

Joseph Baudrillard[1,2]([envelope]), Sébastien Guillon[1], Jean-Marc Chassery[2], Michèle Rombaut[2], and Kai Wang[2]

[1] TOTAL SA, CSTJF, Avenue Larribau, 64000 Pau, France
{joseph.baudrillard,sebastien.guillon}@total.com
[2] Univ. Grenoble Alpes, CNRS, Grenoble INP, GIPSA-lab, 38000 Grenoble, France
{jean-marc.chassery,michele.rombaut,kai.wang}@gipsa-lab.grenoble-inp.fr

Abstract. Oil & gas exploration requires modeling underground geological objects, notably sediment deposition surfaces called "horizons" which are piecewise-explicit surfaces. Starting from sparse interpretation data (e.g. polylines and heightmap fragments), a denser model must be built by interpolation. We propose here a reconstruction method for multivalued horizons with reverse faults. It uses an abstract graph to provide a unified representation of input interpretation data. The graph is partitioned into simpler (explicit) parts which are then jointly interpolated in the plane by a natural extension of standard horizon interpolation methods.

Keywords: Multivalued horizon · Surface reconstruction · Non-explicit surface · Polyline · Heightmap · Interpolation · Gridding · Graph

1 A New Model for Multivalued Horizons

1.1 Limits of Current Horizon Model

Oil & Gas Exploration. In order to locate hydrocarbons deposits, oil & gas companies conduct seismic reflection surveys which are echographies of the underground. This produces seismic cubes, containing a digital image of the subsurface acoustic reflectors. Amongst the objects interpreted on those cubes, sediment deposition surfaces (called horizons) are of first importance. On noisy cube areas, they are picked by geologists as a sparse set of *polylines*. When Signal to Noise Ratio (SNR) is high, automatic methods are instead used to propagate the horizon surface in the cube from seed points, resulting in *heightmap fragments*. In the end, this sparse interpretation data (polylines and heightmap fragments) is interpolated into a denser representation (a complete heightmap) using gridding methods [5,17].

Work performed in a CIFRE PhD thesis in partnership between ANRT, GIPSA-LAB and TOTAL SA.

L. Ragia et al. (Eds.): GISTAM 2018, CCIS 1061, pp. 1–27, 2019.
https://doi.org/10.1007/978-3-030-29948-4_1

From Heightmap to Multivalued Horizons. Heightmaps are digital elevation models, namely a geolocalized image whose pixel values are a vertical elevation distance. This is adequate for representing explicit surfaces, i.e. surfaces with at most one elevation for a given position in the heightmap. This is often valid for horizons, as sediment deposition is a gravity-bound process, leading to flat "monovalued" horizons.

However, horizons can undergo significant geometry change by various natural geological processes. Amongst them is the formation of faults: in compressive domain, rocks under stress eventually crack, forming reverse faults. This leads to the vertical superposition of horizon parts, making the surface not explicit anymore. Several heights would be required to model it with a heightmap, hence such horizons are called "multivalued" (see Fig. 1).

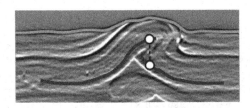

Fig. 1. Monovalued (top) and multivalued (bottom) horizons interpreted on a seismic section. At some heightmap location (corresponding to a vertical dashed line in this section view), 2 points are required to describe the multivalued (bottom) surface. *Figure also found in* [2].

This means multivalued horizons cannot be represented by a single heightmap. This is problematic, as a heightmap has many advantages: it is a 2D object with a compact memory footprint, and fast random access to surface elements is possible (pixels indeed have an implicit location in the heightmap). On practical terms, a lot of software was written with this assumption in mind, preventing the use of a radically different model in TOTAL's exploration platform.

Objectives. Our first objective is therefore to propose a new model for horizons when they are multivalued. This new model must be similar to the old one because of strong performance constraints and software legacy reasons. The second task is to develop a reconstruction method in order to populate this new model by interpolating sparse interpretation data (polylines and heightmap fragments).

1.2   A New Horizon Model: The Patch System

State of the Art. A single heightmap is not enough to represent a multivalued horizon. There are however other numerical models able to represent a multivalued horizon, which can moreover be built by interpolation from sparse geological data.

Point clouds can indeed be used [13]. Triangulated surfaces (meshes) can be constructed from seismic data [16]. Various approaches also exist to interpolate closed meshes from contours but they cannot cope with open surfaces [20]. Alternatively, one could consider using voxels to represent a surface, though this requires using memory efficient spatial index structures such as voxel octrees [14] or R-tree variants [3]. A surface can also be parameterized into an image [10], but knowledge of the complete surface is needed, which is not our case as we interpolate it from sparse polylines and heightmap fragments.

In the next section, we will show that these alternative models perform significantly worse than a heightmap when it comes to representing horizons, and prevent a unified representation of monovalued horizons (with a heightmap) and multivalued horizons. Instead, a new model naturally extending the heightmap to piecewise-explicit surfaces will be used. This new model will also support gridding without major modification.

Performance Comparison. The whole problem takes place in a digital seismic survey of horizontal dimensions $W \times H$ pixels, the horizontal plane being associated to the first two coordinates (x, y) of a 3D point (x, y, z). It is illustrated in Fig. 2. Noting $[\![a, b]\!]$ the set $\{n \in \mathbb{N}, a \leq n, n \leq b\}$, we therefore define the *survey domain* $\mathcal{D}$ such as:

$$\mathcal{D} \doteq [\![0, W - 1]\!] \times [\![0, H - 1]\!] \tag{1}$$

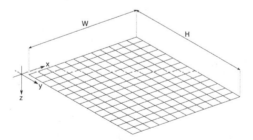

Fig. 2. The survey 2D grid within 3D space. *Figure also found in* [2].

For explicit (monovalued) horizons, the heightmap is a really efficient model: pixels can be accessed in constant time because of their implicit location, and an image is also a very compact data structure in memory. It can easily be displayed as a map, or triangulated into a mesh. In other words it has a simple data model, and can be quickly created, displayed or processed.

Seismic surveys can reach large dimensions (hundreds of gigabytes on disk). A heightmap on this kind of survey can therefore be an image dozens of megapixels large, and hundreds are routinely computed during a study. Moreover, many geophysical processes and methods were developed as image processing algorithms and hence require the regular sampling of a heightmap. Using a radically different model such as a mesh would lead to significant methodology change and software refactoring.

We complemented these qualitative arguments by a benchmark. It compared the performance of all models mentioned in Sect. 1.2 in typical usage situations (IO, display, storage). The quantitative results we gathered confirmed our initial project to keep using an explicit representation (the heightmap) as the model for multivalued horizons[1]. It must however be changed in order to cope with vertically superposed horizon parts that come with multivalued surfaces.

Model Proposal. The proposed extension of heightmaps to multivalued horizons is a set of connected heightmaps, called a *patch system*. The idea is to use as many heightmaps as required: two vertically superposed points must indeed belong to two separate heightmaps. We also want to support connections between pixels of different heightmaps so a complete connected surface can be described (see Fig. 3).

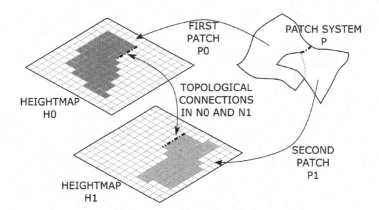

Fig. 3. An example of multivalued surface described by a patch system made up of two patches P_0 and P_1. The geometry is stored in the two heightmaps H_0 and H_1, whereas the per-pixel topological connections between the two patches are stored in the two neighbor data structure N_0 and N_1. *Figure also found in* [2].

More formally, a patch system P made up of N_P patches P_i can be defined as:

$$P \doteq \{P_i, i \in [\![0, N_P - 1]\!]\} \tag{2}$$

---

[1] Detailed numerical results are not shown here for brevity.

Each patch P_i is defined as:

$$P_i \doteq \{H_i, N_i\} \tag{3}$$

where:

- H_i is the patch *heightmap*, storing the geometry of the patch:

$$H_i : \begin{cases} \mathcal{D} & \to \mathbb{R} \\ (x,y) & \mapsto \text{Patch height } z \text{ at } (x,y) \end{cases} \tag{4}$$

- N_i is the so-called *neighbor data structure* containing the neighborhood information. Namely it provides the natural pixel neighbors of any pixel of heightmap H_i, and if necessary the pixel neighbors in another patch $P_j, j \neq i$. The latter only occurs for pixels touching another patch, i.e. on the "edges" of a patch. This structure stores the topology of the patch system, and can typically be constructed as a *neighbor patch index map* that provides for each pixel a list of neighbors. A neighbor in this list is a pair of patch index and neighbor index (for example, from 0 to 3 for the four Von Neumann neighbors). This is made space efficient by omitting neighbors that are in the same patch, in other words neighbors with the same patch index.

It follows that a patch system is a piecewise-explicit representation of a multivalued horizon, and benefits from the efficiency of the heightmap model for both storage and access. Our model elegantly extends the heightmap, and a monovalued horizon can be seen as a patch system with a single patch, without any useless overhead or complexity being introduced. We will demonstrate how standard interpolation algorithms can be adapted to this model in Sect. 1.3, given a properly defined patch system. The construction of such a well-formed patch system from sparse polylines and heightmap fragments is the subject of Sect. 2.

1.3   Horizon Interpolation: The Gridding Process

Monovalued Case. Monovalued horizons are represented by heightmaps. For practical reasons heightmaps are constructed by interpolation of a set of sparse polylines and heightmap fragments, picked or propagated by geologists on the seismic cube. In this context the interpolation process is called *gridding*. Many interpolation methods can be used (inverse-distance weighting, kriging) but we chose a variational approach [5,17] which is straightforward, efficient and robust towards constraint density anisotropy which can be severe in our case.

Objectives: Given a set of constraint height values $f_{i,j}$ to respect at positions (i,j), the objective is to find the unknown heights elsewhere on the heightmap, while creating a smooth surface. This can be formulated as the search for an unknown function $f : \mathcal{D} \mapsto \mathbb{R}$ that takes the values $f_{i,j}$ at locations (i,j) while being smooth.

Variational Formulation: Let Ω be the set of locations where the height is known. Horizon gridding can be seen as a minimization problem of a quantity $J(f)$ defined by two components $D(f)$ and $L(f)$, the first quantifying how close the surface is to the constraint heights, the second measuring the "smoothness" of the final surface:

$$J(f) \doteq D(f) + L(f) \tag{5}$$

Where:

– $D(f)$ is the constraint term that imposes f to pass through known values at locations in Ω. It is defined as:

$$D(f) \doteq ||f \cdot \delta - \overline{f}||^2 \tag{6}$$

With:

- δ being the selection function such as:

$$\delta : \begin{cases} \mathcal{D} & \to \{0,1\} \\ (i,j) & \mapsto \begin{cases} 1 & \text{if } (i,j) \in \Omega^2 \\ 0 & \text{otherwise} \end{cases} \end{cases} \tag{7}$$

- $\overline{f}$ is f evaluated at locations in Ω:

$$\overline{f} : \begin{cases} \mathcal{D} & \to \mathbb{R} \\ (i,j) & \mapsto \begin{cases} f_{i,j} & \text{if } (i,j) \in \Omega^2 \\ 0 & \text{otherwise} \end{cases} \end{cases} \tag{8}$$

– $L(f)$ is the smoothness term. In our case we want to minimize the variations and curvature of f which is expressed using a linear combination of the gradient and Laplace operators, as they provide an image of the local slope and curvature respectively[2]:

$$L(f) \doteq \alpha||\nabla f||^2 + \beta||\Delta f||^2 \tag{9}$$

The Gridding Equation: We want to minimize the quantity $J(f) = ||f \cdot \delta - \overline{f}||^2 + \alpha||\nabla f||^2 + \beta||\Delta f||^2$. This is reached when $\frac{\partial J}{\partial f} = 0$, which leads to:

$$(\alpha\Delta + \beta\Delta^2 + \delta)f = \overline{f} \tag{10}$$

The parameters α and β in Eqs. 9 and 10 can be made small in order to have a surface closer to input constraints. Conversely, big values lead to a very smooth surface that might respect constraints more loosely. Control over uncertainties can therefore be obtained using relevant parameter values.

---

[2] As written, $L(f)$ prevents an exact passage through constraints but this is acceptable in our context.

Numerical Implementation: When evaluated numerically using the finite difference method, by noting $n = W \cdot H$ and by mapping the survey grid on a vector of $\mathbb{R}^n$, it can be shown that this leads to the definition of a matrix equation in the form:

$$A \cdot \mathbf{x} = \mathbf{b} \qquad (11)$$

where:

- A is an $n \times n$ matrix representing the action of operator $(\alpha\Delta + \beta\Delta^2 + \delta)$, i.e. the access to neighbor pixels;
- $\mathbf{x}$ is a vector of $\mathbb{R}^n$ containing the unknown height pixels f;
- $\mathbf{b}$ is a vector of $\mathbb{R}^n$ containing 0 for pixels without constraints, and the known height $\overline{f}$ for constraint pixels.

Monovalued gridding is performed by first rasterizing the polylines onto the heightmap using standard algorithms [4]. Pixel positions and heights are interpolated between polyline vertices in this process. As for heightmap fragments, they are projected on the complete heightmap. Equation 11 can then be solved by a direct or iterative method (Jacobi, Gauss-Seidel, conjugate gradient, etc.). Implementations are presented in details in the literature [19].

Figure 4 illustrates the gridding of some projected polylines and a heightmap fragment. Note that not all pixels of the heightmap become valued (the surface does not take all the image). Indeed, gridding only takes place in what we call the *envelope* of the horizon, i.e. the pixel locations where it should be defined. Outside envelope, extrapolation would occur instead of interpolation[3].

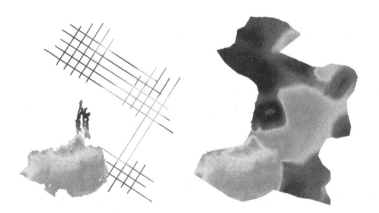

Fig. 4. An example of monovalued gridding shown on a map (i.e. viewed from top). A color ramp is used to represent elevation. Sparse constraint pixels from rasterized polylines and heightmap fragments (left) are interpolated into a dense surface (right).

---

[3] The actual definition of an envelope for monovalued horizons and how it prevents pixels from being gridded are not detailed here for the sake of brevity.

Multivalued Case. When looking at Eq. 11, we see that only the connectivity information stored in A depends on the horizon type: it is a simple access to natural pixel neighbors in the case of a monovalued horizon, and becomes slightly more complex in the case of a multivalued horizon – a patch pixel can have neighbors in another patch. In order to grid multivalued horizons, we define an extended unknown vector $\mathbf{x_E}$ that contains the unknown heights $\mathbf{x_i}$ associated to all patches P_i simultaneously:

$$\mathbf{x_E} \doteq \begin{pmatrix} \mathbf{x_1} \\ \vdots \\ \mathbf{x_N} \end{pmatrix} \tag{12}$$

Using the connectivity information given by the neighbor data structure presented in Sect. 1.2, it is possible to construct the extended operator matrix A_E and the extended constraint vector $\mathbf{b_E}$ in a similar manner and therefore interpolate each patch correctly. Equation 11 then becomes:

$$A_E \cdot \mathbf{x_E} = \mathbf{b_E} \tag{13}$$

Interpolation of a multivalued horizon in the form of a patch system is then a natural extension of monovalued gridding. However, given just a set of input polylines and heightmap fragments, in order to prepare a patch system for gridding, we must first convert the input interpretation data into a unified graph (Sect. 2.1), then provide a solution to a partitioning problem (Sect. 2.2) and finally solve an envelope computation problem (Sect. 2.3).

2   Multivalued Gridding Preparation

2.1   A Graph View of Input Interpretation Data

Input Polylines. Recall geologists pick polylines in the low SNR areas of the seismic cube. An example of such a polyline is illustrated by Fig. 5. We can therefore define p, the set of N_p input polylines, as:

$$p \doteq \{p_i, i \in [\![0, N_p - 1]\!]\} \tag{14}$$

– Each polyline p_i can be written as follow:

$$p_i \doteq \{V_j = (x_j, y_j, z_j), j \in [\![0, N_{p_i} - 1]\!]\} \tag{15}$$

i.e. it is a set of N_{p_i} 3D points. Those points are scattered within the survey. They are considered to form an open polyline, each being connected by an edge to the previous and next vertices in the set p_i – except for the first and last vertices;

– Each polyline vertex is placed along a regular grid in the first two dimensions, as it belongs to the survey of size $W \times H$ pixels. The third axis use real numbers for better vertical precision. This can be summarized as:

$$\forall i \in [\![0, N_p - 1]\!], \forall (x, y, z) \in p_i, \begin{cases} x \in [\![0, W - 1]\!] \\ y \in [\![0, H - 1]\!] \\ z \in \mathbb{R} \end{cases} \tag{16}$$

– Note we use uppercase P for the set of patches P_i in a patch system, while lowercase p is the set of input polylines p_i.

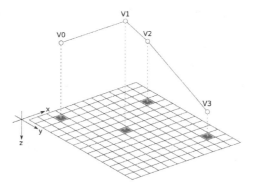

Fig. 5. A polyline made up of 4 vertices, and their horizontal location on the survey grid. *Figure also found in* [2].

Input Heightmap Fragments. In high SNR regions of the seismic cube, automatic picking of horizons can be performed (for example by propagation from seed seismic samples [11]). This leads to the creation of heightmap fragments in parts of the survey. We define h, a set of N_h heightmaps such as:

$$h \doteq \{h_i, i \in [\![0, N_h - 1]\!]\} \tag{17}$$

– Each heightmap is an image of the same size than the survey:

$$h_i : \begin{cases} \mathcal{D} & \to \mathbb{R} \\ (x, y) & \mapsto h_i(x, y) = \begin{cases} z & \text{if pixel is valued} \\ \gamma & \text{(outside surface) otherwise} \end{cases} \end{cases} \tag{18}$$

– γ (outside surface) is the magic pixel value meaning that at this location, the heightmap is not defined (e.g. it is outside of or a hole in the surface);
– It is therefore possible for heightmaps to have holes in them, associated with a zone of one or more γ valued pixels. It means that the heightmap pixels at the boundary of such holes have less than 4 valued neighbor pixels;

- By definition each heightmap, hence each heightmap connected component, describes a monovalued (explicit) surface;
- Once again, note we use uppercase H for the heightmaps H_i of a patch system, while lowercase h is the set of input heightmap fragments h_i, leading to the input heightmap connected components $h_{i,j}$ (see below).

Each heightmap h_i can be made up of several connected components $h_{i,j}$ as depicted Fig. 6. As for polylines (each being a single connected component), we will rather more consider the set h^* of heightmap connected components, instead of the set of heightmaps h. We can now grant that each object in h^* is, by construction, made up of a single connected component, i.e. from any pixel we can reach any other. This is not incompatible with holes but forbids recursive inclusion of valued pixels inside a hole, for example as in Fig. 6.

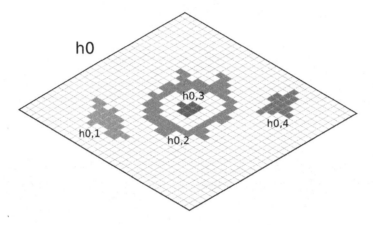

Fig. 6. This heightmap h_0 is made up of 4 connected components. It cannot be used for our process however, because there is a recursive inclusion of pixels in a hole, i.e. $h_{0,3}$ is within $h_{0,2}$.

An Abstract Graph of Interpretation Data. An elegant and practical way to handle both polylines and heightmap fragments as unified inputs for our reconstruction scheme is to see the input data as forming an abstract graph G (see Fig. 7), that will be constructed in two steps:

- A *geometry pass* to create the set of graph vertices V, representing input interpretation data (be it a polyline or a heightmap connected component);
- A *topology pass* to create the set of graph edges E, associated with a topological link between two interpretation data instances (for example, two polylines intersecting or a polyline ending on a heightmap).

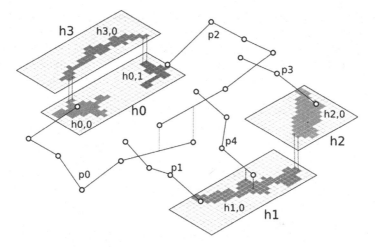

Fig. 7. Input graph of sparse interpretation data (polylines and heightmap fragments, each being potentially made up several connected components).

Geometry Pass: Creating Graph Vertices. The first step is to gather both polylines in p and heightmap connected components in h^* into a single set of input interpretation data, namely V, that will be the vertices of the abstract graph G. By construction, each element in V will be either a polyline or a heightmap connected component, i.e. it will be monovalued (explicit) and made up of a single connected component. Figure 8 provides an example of such abstract graph vertices, coming from the input data in Fig. 7.

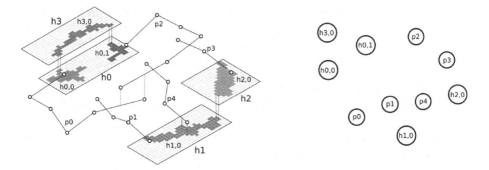

Fig. 8. An abstract graph representing interpretation data (either heightmap connected component or polyline) in its vertices. Junctions (topological connections between interpretation data) will later be represented with graph edges.

Topology Pass: Creating Graph Edges. The second step is to detect the junctions between any two interpretation data, and create a graph edge in E for each junction. This also leads to the possible splitting of graph vertices (for example two polylines snapped in the middle lead to four polylines connected at an end). This is done in order to get connections only at the border of interpretation data, be it a polyline end vertex or a heightmap connected component's border pixel. Two graph vertices will be moreover added in order to represent the two new polylines. Because we have two classes of interpretation data (polyline or heightmap connected component), we have three possible junctions:

- Polyline-Polyline junction (PP). This happens when two edges coming from two distinct polylines are very close – recall polylines are hand-picked on computer screens, so they cannot intersect exactly numerically speaking. Given a so-called snapping threshold d_S, we can therefore snap together the polyline edges and introduce a common polyline vertex, as illustrated in Fig. 9;

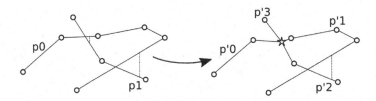

Fig. 9. Close edges from different polylines are snapped at a common polyline vertex. Edges farther than d_S are not snapped and stay superposed. This leads to four polylines, connected together at an end vertex represented by a five branched star.

- Heightmap-Heightmap junction (HH). Automatic propagation of heightmap fragments from several starting points can lead to superpositions between two or more heightmap connected components. By a process similar to polyline snapping, we can join two superposed heightmap connected components along a boundary curve made of pixels (see Fig. 10)[4];
- Polyline-Heightmap junction (PH). Polylines are often picked in low SNR areas, whereas heightmaps can be propagated in high SNR zones. This means they provide two complementary types of input interpretation data, that must be eventually connected together. As for PP and HH junctions, a polyline passing close enough to a heightmap connected component will lead to a connection between a border pixel and a polyline vertex, as shown in Fig. 11.

We now have a graph G of interpretation data, whose vertices are interpretation data geometry (polyline or heightmap connected component). The graph vertices have an "intra" topology (polyline edges joined by polyline vertices, or implicit connectivity between valued pixels of heightmap connected

---

[4] This junction process is not detailed here but is in the spirit of the dilated envelope restriction in Sect. 2.3.

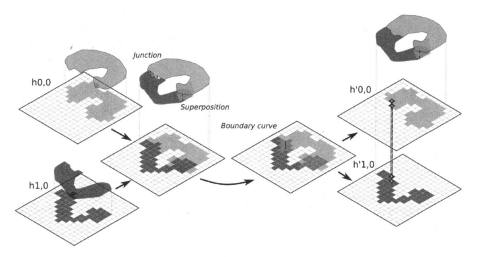

Fig. 10. Two heightmap connected components that have both superposed parts and "almost joined" parts. Superposed parts are shown in red and are too far away to be joined. Close enough parts (in green) are made to nicely join each other along a boundary curve made of pixels, represented by a four branched star. (Color figure online)

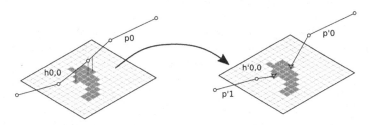

Fig. 11. A polyline close to a heightmap is snapped on two border pixels, represented by a three branched star.

components), but are by construction monovalued and made up of a single connected component. Figure 12 provides an example of the graph at this stage. The graph edges represent "inter" topology, arising in the three situations previously described. At this point the graph G can be partitioned into monovalued sub-graphs that we will interpolate.

For the rest of our workflow, We can assume without loss of generality that G is a connected graph, i.e. all input interpretation data was picked to represent a single connected surface. If not, each connected component can be handled independently as a separate connected graph.

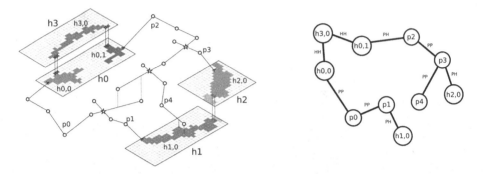

Fig. 12. The abstract graph after junctions are handled: there are now graph edges representing topological connections between input interpretation data. Stars with 3, 4 and 5 branches represent PH, HH and PP junctions respectively. The abstract graph is now connected.

2.2 Partitioning Problem

The objective is now to find a decomposition of G into a set of monovalued sub-graphs G_i, i.e. sub-graphs where no internal vertical overlap occurs. Such decomposition is not unique, therefore some criteria must be defined in order to choose a suitable partition.

After a partition is found, each sub-graph will be turned into a patch heightmap in the envelope computation stage, and will then be gridded. Both these steps have a computational complexity of $\mathcal{O}(N \cdot W \cdot H)$ where N is the number of patch in the patch system, W and H are the width and height of the patch. This means we want to reduce the number of patch (so the number of sub-graphs G_i) as much as possible, and large patch size must be avoided – this is a second order concern though as only the envelope will be considered, not the entire heightmap image.

Considering the partitioning problem in a variational framework could provide an optimal combination of patch count and size, but would be prohibitive to evaluate, graph partitioning problems often being NP-hard [6]. In this context, we propose instead a constructive method that leads to an acceptable compromise between patch count and size. The approach is based on three steps:

- **Multivalued Scan.** Vertically superposed interpretation data of G are detected, grouped into superposed zones, and graph vertices are introduced in order to avoid having half-superposed interpretation data;
- **Sub-graph Index Propagation.** Simultaneous propagation of sub-graph index from superposed zones leads to the definition of monovalued sub-graphs G_i;
- **Merge.** Reduce sub-graph count by merging together those that can be. The sub-graphs after merge are noted $\tilde{G}_i$.

Multivalued Scan: Detecting Superpositions in the Graph. The objective of this section is first to detect when the interpretation data associated with two graph vertices are vertically superposed, and second to split them such that any two graph vertices are either completely superposed or not at all. As for the three possible junction situations, there are three different kind of superposition:

- Polyline-Polyline superposition (PP). Polylines are picked in planar sections of the cube (vertical, horizontal or arbitrary). This means some polyline edges can be vertically superposed. There is no reason for two edges to be entirely superposed though, so we introduce a polyline vertex whenever necessary so that a polyline edge is either completely superposed with another, or is not at all (see Fig. 13);
- Heightmap-Heightmap superposition (HH). Some heightmap connected components can vertically overlap, while having a significant distance between them: they won't be joined as in Sect. 2.1. Instead, each must be split into several heightmap connected components such that each is either completely superposed with another, or not at all. This is illustrated in Fig. 14;
- Polyline-Heightmap superposition (PH). By seeing a polyline as a heigtmap (for example by rasterization [4]), a PH superposition is nothing but a degenerate HH superposition and can be solved as shown previously (see Fig. 15).

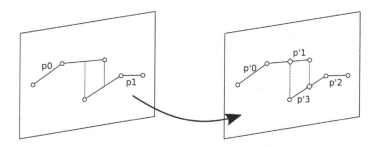

Fig. 13. Polyline vertices (symbolized here by diamonds) are introduced and polylines split in order to only have polylines that are totally overlapping, or not at all.

Whatever the superposition type, additional graph vertices will be added in order to enforce total or zero superposition for any two graph vertices. Once these vertices are introduced, detecting vertically superposed graph vertices is a simple geometric problem. Figure 16 shows the final graph after superpositions are handled.

Sub-graph Index Propagation. At this point, we can give an index i for each graph vertex that is superposed, and for each we initialize a *monovalued sub-graph* G_i with the graph vertex. The sub-graphs G_i are called monovalued as by construction, each is made of graph vertices that do not overlap, each

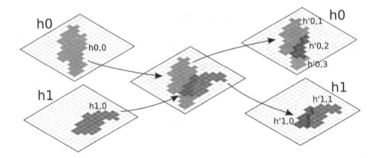

Fig. 14. Superposed heightmap pixels lead to three new heightmap connected components, joining the previous ones along boundary curves.

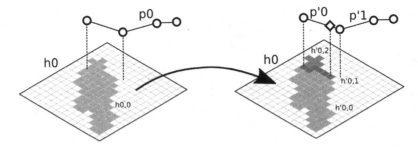

Fig. 15. A polyline superposed with a heightmap, leading to the creation of a new heightmap along superposed pixels. The polyline must also be split in two and a polyline vertex (drawn as a diamond) is introduced.

one being also monovalued by construction. All the graph vertices can then be indexed using a propagation method illustrated by Algorithm 1.

This ensures that all graph vertices will have a sub-graph index, but more importantly that each index will be associated with a similar number of graph vertices[5]. Following the previous example, Fig. 17 shows the evolution of the sub-graph index propagation in the graph.

Merge. By starting from superposed graph vertices, sub-graph index propagation ensures that enough monovalued sub-graphs will be used. However it can lead to a massive over-estimation of the number of required sub-graphs, especially when the input interpretation data is dense. This being said, it occurs that many of the sub-graphs can be merged together.

Indeed, let us consider a pair of sub-graphs $(G_i, G_j), i \neq j$. If they are connected by a graph edge and do not have graph vertices that overlap vertically,

---

[5] Exact same number is not reached as it depends on the graph shape for propagation. Moreover, two graph vertices can have a different geometrical extent, e.g. polylines being smaller than heightmap connected components.

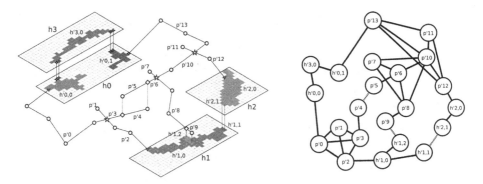

Fig. 16. Abstract graph after superpositions were detected. Extra graph vertices were added as described previously, and superpositions between graph vertices are noted with a dashed line.

then they are merged together. An example of merging can be found in Fig. 17. We call $\tilde{G}_i$ the merged sub-graphs.

2.3   Envelope Computation

At this point the partitioning problem is solved as we found a partition of G into a relatively low number of monovalued sub-graphs $\tilde{G}_i$. In order for a sub-graph $\tilde{G}_i$ to be gridded, it is however necessary to convert it to a patch P_i and compute its envelope. As detailed in the next section, each sub-graph will indeed be converted into a heightmap, and its polylines and heightmap connected components will be rasterized into *constraint pixels*.

For a patch P_i, the envelope is the combination of two objects:

– A *mask* indicating for each pixel of its heightmap H_i whether it is to be gridded or not. This mask will be encoded in the heightmap H_i using a boolean value, for example *true* if inside envelopes, *false* otherwise;
– A set of *junction points*, i.e. pixels that have neighbor pixels in another patch. This will be encoded in the neighbor data structure N_i.

The envelope will therefore be the domain around constraint pixels, i.e. pixels coming from interpretation data. There are methods to compute the envelope (or "hull") of a set of pixels: one could consider using the pixels' convex shape [12] or alpha shape [8], but in our case this lead to masks that are too large and hence does not prevent extrapolation.

An efficient and intuitive way to construct this mask is instead to use the closing morphological operator against the constraint pixels of each patch heightmap. Closing is actually the succession of a dilatation and an erosion, both using a structural element of size $d_C \in \mathbb{N}^*$ pixels. When a relevant value of d_C is chosen, holes between constraint pixels are closed by the dilatation while extrapolation is avoided because of the erosion. We therefore propose the following steps to find the envelope of each patch:

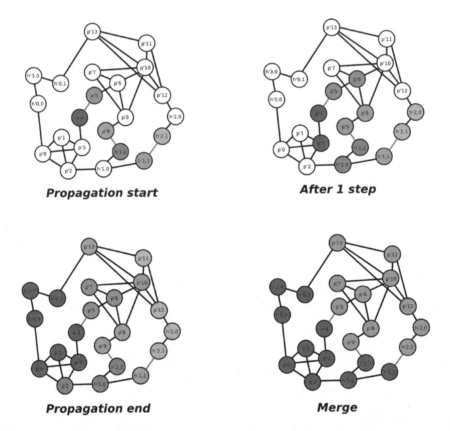

Fig. 17. Before propagation, only superposed graph vertices are given an index, symbolized here by a color. At each propagation step, each color gets propagated in the graph until all graph vertices have a color. After propagation end, sub-graphs (i.e. graph vertices of the same color) are merged together to reduce the sub-graph count, as explained in the following section.

- **Heigtmap Conversion.** Turn each sub-graph $\tilde{G}_i$ into a patch heightmap H_i initialized with constraint pixels;
- **Dilatation.** A dilated envelope is created independently for each patch;
- **Dilated Envelope Restriction and Junction Point Location.** For each patch pair that is connected topologically by an edge of G, restrict dilated envelopes to ensure smooth connection along a set of junction points;
- **Joint Erosion.** Each dilated envelope is eroded to prevent extrapolation. This is done simultaneously, i.e. on the multivalued surface.

Heightmap Conversion. Each sub-graph can now be converted into an image of size $W \times H$, the survey size (see Fig. 18). We therefore associate each sub-graph $\tilde{G}_i$ with its corresponding patch P_i of heightmap H_i whose pixel (x, y)

Algorithm 1. Sub-graph index propagation.

   Procedure propagate $(G, \{G_i\})$
   Input:
      G ▷ Connected abstract graph
      $\{G_i\}$ ▷ Indexed sub-graphs (superposed graph vertices only at start)
   Algorithm:
1: $vertices \leftarrow$ FIFO list with all vertices of $\{G_i\}$
2: **while** $vertices$ is not empty **do**
3:     Pop a, the first vertex of $vertices$
4:     **for** Each unindexed vertex b touching a **do**
5:        $index \leftarrow a$'s index
6:        Index b with $index$
7:        Add b to $vertices$
8:     **end for**
9:     Add a to G_{index}
10: **end while**
   End procedure

---

contains the height z for any interpretation data (x, y, z) in $\tilde{G}_i$:

$$
H_i : \begin{cases} \mathcal{D} & \to \mathbb{R} \\ (x, y) & \mapsto \begin{cases} z' & \text{if } \exists M' = (x', y', z') \in \tilde{G}_i, \\ & \quad (x, y) = (x', y') \\ \nu & \text{(null value) otherwise} \end{cases} \end{cases} \tag{19}
$$

Remarks:

- ν (null value) is a magic value designating a patch pixel that is not yet valued (it is not a constraint pixel). The value of such a pixel will be set during gridding;
- Concretely, H_i is obtained by rasterizing any polyline in $\tilde{G}_i$ and projecting any heightmap connected component in $\tilde{G}_i$;
- The "intra-patch" connectivity information once stored explicitly in the vertices and edges of $\tilde{G}_i$ is now replaced by the natural neighborhood of the pixels in P_i. The "inter-patch" connectivity, i.e. the topological connection between $\tilde{G}_i$ and its potential neighbor sub-graphs is for now lost though, but it will be stored in the neighbor data structure N_i when computing the dilated envelope restriction and the joint erosion.

Dilatation. Although image morphological operators are typically defined by kernels associated with structural elements, numerical implementations are faster when using Euclidean Distance Maps (EDM). It can be shown that both dilatation and erosion are equivalent to the thresholding of an EDM[6] [15]. Using an

---

[6] This is for disk-shaped structural elements and distance maps based on the $L2$ norm, because the disk is the topological ball associated with the $L2$ norm in $\mathbb{R}^2$.

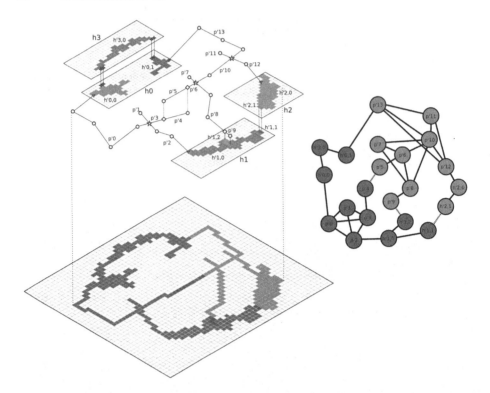

Fig. 18. Following the example in Fig. 17, each subgraph G_i is converted into a heightmap H_i. The superposed result is displayed here. Overlapping graph vertices in G lead to overlapping pixels in this image. Polylines are rasterized and heightmap connected components are projected in order to get this raster representation.

EDM is faster than using masks because there are efficient $\mathcal{O}(W \cdot H)$ algorithms to compute an approximated distance map [7,18]. Using approximations is tolerable in our case as the envelope does not require pixel-perfect precision and those errors are small [9].

Recall constraints are the non-ν pixels of heightmap H_i. We therefore construct the map of *distance to constraints* DC_i. Being a distance map, each pixel of DC_i has a positive value and is only zero on the location of constraints, i.e. ν pixels. DC_i is defined by:

$$DC_i : \begin{cases} \mathcal{D} & \to \mathbb{R}^+ \\ (x, y) & \mapsto \text{distance to closest non-}\nu\text{ pixel} \end{cases} \tag{20}$$

We then define the *dilated envelope* DE_i by thresholding the distance map DC_i:

$$DE_i \doteq \{(x, y) \in \mathcal{D}, DC_i(x, y) \leq d_C\} \tag{21}$$

Using this thresholding method, it is possible to obtain the dilated envelopes, as depicted by Fig. 19.

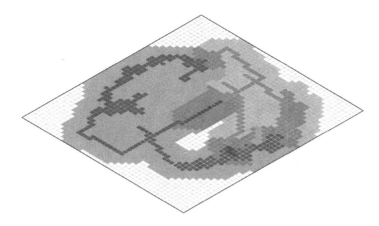

Fig. 19. An example of dilated envelopes. They overlap when constraints of the two patches are both closer than d_C.

Dilated Envelope Restriction and Junction Point Location. Before computing the erosion, we want dilated envelopes to join along a *boundary curve* without overlapping around the known topological connections between two patches, i.e. near edges of G that have vertices from two sub-graphs (G_i, G_j), $i \neq j$. Meanwhile, we also want to allow and preserve dilated envelopes overlapping around superposed constraints (for example in Fig. 19, superposed polyline edges must eventually lead to superposed parts of the final multivalued surface). This can be handled simultaneously by a criteria map using the following procedure.

Compute Criteria Map: Each boundary between two patches P_i and P_j should be located "in the middle" of the two patches' dilated envelopes. For this reason we compute a *criteria map* $C_{i,j}$ derived from distance maps DC_i and DC_j: see Fig. 20 for an example. The criteria map can be defined as:

$$C_{i,j} : \begin{cases} \mathcal{D} & \to \mathbb{R} \\ (x,y) & \mapsto DC_i(x,y) - DC_j(x,y) \end{cases} \qquad (22)$$

Remarks:

- A pixel in criteria map $C_{i,j}$ has negative value when closer to patch i than patch j;
- A pixel in criteria map $C_{i,j}$ has positive value when closer to patch j than patch i;
- We want the boundary curve between patches i and j to be defined the location of sign change in $C_{i,j}$;
- However, the boundary curve should not be defined around superposed constraints, i.e. on pixels valued 0.

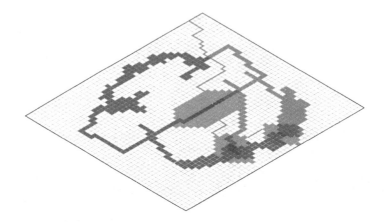

Fig. 20. In green is depicted the isovalue 0 in the criteria map used in order to define the boundary shape. It is "between" the pixels unless on the "0 areas" associated with superposed constraint pixels, where the boundary curve should not be defined. (Color figure online)

Restrict Envelopes: In order to restrict the dilated envelopes, we introduce the set of locations where the criteria map is positive and negative (excluding locations where criteria is zero):

$$\begin{cases} C_{i,j}^{+} \doteq \{(x,y) \in \mathcal{D}, C_{i,j}(x,y) > 0\} \\ C_{i,j}^{-} \doteq \{(x,y) \in \mathcal{D}, C_{i,j}(x,y) < 0\} \end{cases} \tag{23}$$

We then define the *restricted dilated envelopes* RDE_i and RDE_j of patches i and j as depicted by Figs. 21 and 22: the restricted dilated envelope of patch i is the dilated envelope of patch i, but deprived of areas where the criteria map $C_{i,j}$ is strictly positive, i.e. when closer to patch j. "0 areas" are kept on the restricted dilated envelope so superposed envelopes can exist. This can be noted as:

$$\begin{cases} RDE_i \doteq DE_i \setminus C_{i,j}^{+} \\ RDE_j \doteq DE_j \setminus C_{i,j}^{-} \end{cases} \tag{24}$$

Locate Junction Points: Once restricted, the dilated envelopes will perfectly join at the boundary. At this point, the neighbor data structure N_i of each patch P_i can be updated as locally around the boundary, the per-pixel connections between any two patches are known.

Joint Erosion. Whereas dilatation could be computed independently for each patch in previous section, it is required to consider the patch system as a whole during erosion. Once again, using kernel-based morphological operators works but is extremely slow. Using EDM thresholding still speeds up the process, but

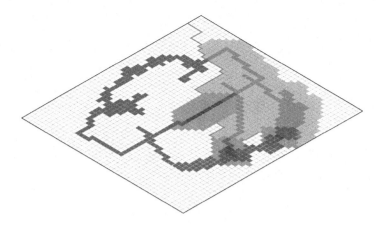

Fig. 21. By keeping "0 areas" while removing envelope beyond the location of sign change in the criteria map, it is possible to define the restricted dilated envelope, here for the right patch as an example.

the fast two-pass algorithm previously used [7] cannot be easily adapted to a non-manifold support, in our case the patch system.

Instead we propose to use a fast-marching algorithm [18] that propagates pixel by pixel the *distance from outside* the restricted dilated envelope on a "multivalued EDM" DO_i, i.e. a patch system whose heightmaps are EDM. As our patch system model clearly defines neighborhood relations in the entire horizon using the neighbor data structures N_i, fast-marching implementation is straightforward.

Once computed, the multivalued EDM DO_i can be thresholded using the closing distance d_C, leading to the definition of the *eroded envelope* EE_i. The erosion of the example patches is shown in Figs. 22 and 23, depicting the envelopes respectively before and after erosion.

The eroded envelope EE_i is therefore:

$$EE_i \doteq \{(x, y) \in \mathcal{D}, DO_i(x, y) \geq d_C\} \tag{25}$$

Along with restricted dilated envelopes, boundaries are also eroded. The neighbor data structures N_i needs therefore to be updated again at this point to only connect together points that are still on the envelope. By construction, we now have an eroded envelope EE_i for each patch P_i, and all eroded envelopes join nicely along the eroded boundaries.

It is now time to update the patch heightmaps with the envelope information: from now on, each pixel of H_i outside of the eroded envelope EE_i is associated with a boolean value *false* (outside patch) in the envelope mask. At this point the patch system is ready for gridding as described in Sect. 1.3.

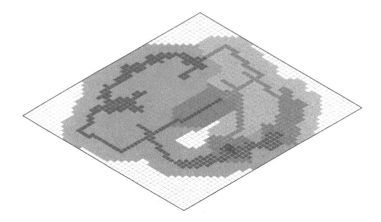

Fig. 22. Restricted dilated envelopes before erosion. Notice the areas "outside" constraint pixels where extrapolation would occur if no erosion was performed.

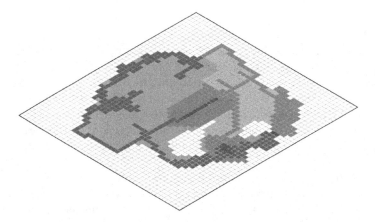

Fig. 23. Cut envelopes after erosion. They still connect along a neat boundary curve, but erosion removed envelope "outside" where extrapolation would have occurred.

3   Results

Examples of reconstructed surfaces using real data are presented here to illustrate the action of our method on polylines and heightmap fragments interpreted by geologists.

The proposed model of patch system as well as the multivalued gridding approach we developed provide good results on both synthetic and real data. We will comment here the interpolation of sparse polylines from a real seismic survey into a patch system[7].

Figure 24 shows the input polylines as picked by geologists on the survey. Many superposed zones will be detected in the multivalued scan, leading to

---

[7] Data size and density will be kept low for readability though.

Fig. 24. Example of real data: sparse polylines and heightmap fragments (left). They are attributed a patch index after abstract graph preparation and partitioning (right).

the indexation of a lot of sub-graphs in the index propagation stage. The final sub-graph count will not be excessive though, because of the merge step.

After conversion to heightmaps, envelope computation begins. This will lead to the definition of valid envelope masks and neighbor data structures, used by the gridding process. The resulting patch system is depicted in Fig. 25. The global surface is smooth even along patch boundaries.

Fig. 25. Following envelope computation and gridding, a patch system is created (left). Patches join smoothly along a boundary curve (right, with another viewing angle and a color map indicating elevation).

4   Conclusion

This multivalued gridding method is an extension of existing work on the reconstruction of horizons from polylines [2]. We showed how to modify the previous workflow in order to take heightmap fragments into account.

An interesting feature would be the handling of faults during the gridding stage. Faults are the result of mechanical failure within a geological object. These discontinuities can displace rock formations on a wide range of distances, some largely visible even at the seismic scale. Horizons can be for example cut and displaced by faults – normal faults being one of the primary source of multivalued horizons. For this reason it makes sense to prevent access to neighbor pixels on the opposite side of a fault while gridding. This is a standard feature of current monovalued gridding implementations in modern geophysics software, and would be appreciated for multivalued horizons as well.

Other types of horizon could also be tackled. Salt domes can for example lead to multivalued surfaces, and current workflow cannot be applied. We proposed another method to reconstruct them [1], but a unified approach would be more elegant.

Beyond new features, many optimizations could also be conducted in order to reduce the run-time and memory footprint of the algorithmic chain. From multi-grid schemes, multi-threading and compression strategies to constraints for the sub-graph index propagation, a lot of progress can be made to support ever larger horizons.

References

1. Baudrillard, J., Guillon, S., Chassery, J.M., Rombaut, M., Wang, K.: Parameterization of polylines in unorganized cross-sections and open surface reconstruction. In: 9th International Conference on Curves and Surfaces, Arcachon, June 2018
2. Baudrillard, J., Guillon, S., Chassery, J.M., Rombaut, M., Wang, K.: Reconstruction of piecewise-explicit surfaces from three-dimensional polylines. In: International Conference on Geographical Information Systems Theory, Applications and Management (Doctoral Consortium), Funchal, March 2018
3. Beckmann, N., Begel, H.P., Schneider, R., Seeger, B.: The R*-Tree: an efficient and robust access method for points and rectangles. ACM Trans. Graph. **19**(2), 322–331 (1990)
4. Bresenham, J.E.: Algorithm for computer control of a digital plotter. IBM Syst. J. **4**(1), 25–30 (1965)
5. Briggs, I.C.: Machine contouring using minimum curvature. Geophysics **39**, 39–48 (1974)
6. Buluç, A., Meyerhenke, H., Safro, I., Sanders, P., Schulz, C.: Recent advances in graph partitioning. In: Kliemann, L., Sanders, P. (eds.) Algorithm Engineering. LNCS, vol. 9220, pp. 117–158. Springer, Cham (2016). https://doi.org/10.1007/978-3-319-49487-6_4
7. Danielsson, E.: Euclidian distance mapping. Comput. Graph. Image Process. **14**, 227–248 (1980)
8. Edelsbrunner, H., Kirkpatrick, D.G., Seidel, R.: On the shape of a set of points in the plane. IEEE Trans. Inf. Theory **29**, 551–559 (1983)
9. Grevera, G.J.: Distance transform algorithms and their implementation and evaluation. Technical report, Saint Joseph's University (2004)
10. Hormann, K., Levy, B., Sheffer, A.: Mesh Parameterization: Theory and Practice. Siggraph Course Notes (2007)
11. Keskes, N., Boulanouar, Y., Zaccagnino, P.: Image analysis techniques for seismic data. SEG Tech. Program Expanded Abstract **2**(1), 221–222 (1982)
12. Kirkpatrick, D.G., Seidel, R.: The ultimate planar convex hull algorithm. SIAM J. Comput. **15**, 287–299 (1986)
13. Levoy, M., Whitted, T.: The use of points as a display primitive. University of North Carolina at Chapel Hill, Computer Science Department, Technical report (1985)
14. Meagher, D.J.R.: Geometric modeling using octree-encoding. Comput. Graph. Image Process. **19**, 129–147 (1982)
15. Russ, J.C.: The Image Processing Handbook. CRC Press, Boca Raton (1998)

16. Sadri, B., Singh, K.: Flow-complex-based shape reconstruction from 3D curves. ACM Trans. Graph. **33**(2), 20:1–20:15 (2014)
17. Smith, W.H.F., Wessel, P.: Gridding with continuous curvature splines in tension. Geophysics **55**, 293–305 (1990)
18. Treister, E., Haber, E.: A fast marching algorithm for the factored Eikonal equation. J. Comput. Phys. **324**(C), 210–225 (2016)
19. Walter, É.: Numerical Methods and Optimization. Springer, Cham (2014). https:// doi.org/10.1007/978-3-319-07671-3
20. Zou, M., Holloway, M., Carr, N., Ju, T.: Topology-constrained surface reconstruction from cross-sections. ACM Trans. Graph. **34**(4), 128:1–128:10 (2015)

Comparison of Traditional and Constrained Recursive Clustering Approaches for Generating Optimal Census Block Group Clusters

Damon Gwinn[1], Jordan Helmick[2], Natasha Kholgade Banerjee[1], and Sean Banerjee[1(✉)]

[1] Clarkson University, Potsdam, NY, USA
{gwinndr,nbanerje,sbanerje}@clarkson.edu
[2] MedExpress, Morgantown, WV, USA
jordan.helmick@medexpress.com

Abstract. Census block groups are used in location selection to determine the average drive time for all residents within a given radius to a proposed new store. The United States census uses 220,334 block groups, however the spatial distance between neighboring block groups in densely populated areas is small enough to cluster multiple block groups into a single unit. In this paper, we evaluate the efficiency and accuracy of drive time computations performed on clusters generated by our novel approach of constrained recursive reclustering as run on three traditional clustering algorithms—affinity propagation, k-means, and mean shift. We perform comparisons of our constrained recursive reclustering approach against drive times computed using the original census block group, and using clusters obtained by traditional reclustering. Unlike traditional clustering, where clustering is performed in a single pass, our approach continues reclustering each new cluster until a user specified stopping criteria is reached. We show that traditional clustering techniques generate sub-optimal clusters, with large spatial distances between the cluster centroid and cluster points making them unusable for computing drive times. Our approach provides reductions of 81.2%, 83.4%, and 10.2% for affinity propagation, k-means, and mean shift respectively when run on 220,334 census block groups. Using 200 randomly sampled locations each from Lowe's, CVS, and Walmart, we show that compared to the original block groups there is no statistically significant difference in drive time computations when using clusters generated by constrained recursive reclustering with affinity propagation for any of the three businesses, and with k-means for CVS and Walmart. While statistically significant differences are obtained with k-means for Lowe's and with mean shift for all three businesses, the differences are negligible, with the mean difference for each location set being within 30 s.

Keywords: Location selection · Census block group · Affinity propagation · k-means · Mean shift · Constrained · Recursive · Clustering

© Springer Nature Switzerland AG 2019
L. Ragia et al. (Eds.): GISTAM 2018, CCIS 1061, pp. 28–54, 2019.
https://doi.org/10.1007/978-3-030-29948-4_2

1   Introduction

Location selection is used to analyze the feasibility of a new location, such as a new medical facility or retail store, by comparing the number of potential customers and their average drive time to those of competing locations. A new location is less likely to succeed if it is too distant from the customer base or is out-positioned by a competitor. Census block groups are used to compute drive times for potential customers, as computing drive times for each individual resident is infeasible. The United States 2010 Census utilizes 220,334 block groups, with each block group representing between 600 and 3,000 people [9].

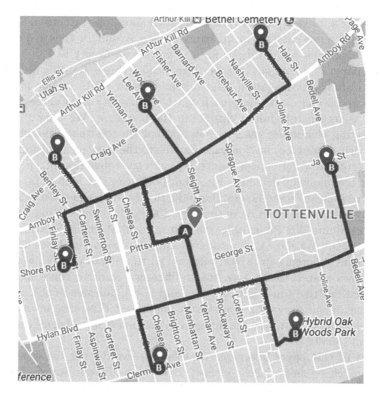

Fig. 1. Tottenville, Staten Island, NY is described by 8 block groups. The distance from the block group labeled A to all block groups labeled B is 1 mile, with a drive time between 2–3 min.

To determine the feasibility of a given location, companies need to evaluate thousands of potential locations and their associated competitors. With major retailers opening hundreds of new locations annually, computing drive times for each potential location from census block groups can become computationally infeasible. For example, Walmart has opened on average 443 new locations worldwide every year for the past 10 years [33]. In Gwinn et al. [18], we presented a constrained recursive cluster splitting technique for generating census block group clusters using affinity propagation as the clustering algorithm. We provided a 5× speed up in the drive time computation process by reducing the number of block groups from 220,334 to 41,442 clustered block groups. Our approach showed no statistically significant difference in drive times when evaluated on a sample of 200 random geographic locations placed within the continental United States and 300 random Walmart locations across the United States. Our recursive reclustering technique is based on the fact that neighboring block groups in densely populated areas show small spatial distances. As shown in Fig. 1, densely populated areas, such as Tottenville, Staten Island, NY, are comprised of multiple block groups. In the figure, block groups labeled B are at a distance of 1 mile, with a drive time between 2 to 3 min, from the block group labeled A. Our approach provides computational speed ups by considering block group A to be representative of all block groups A and B, thus reducing the number of drive time computations for a new location from 8 to 1.

In this paper, we compare the performance of three traditional clustering techniques—affinity propagation, k-means and mean shift—to our novel constrained recursive reclustering approach which can use any of the three clustering algorithms as to generate initial clusters in the base case. We show that traditional clustering techniques generate sub-optimal clusters that are unusable for computing drive times for location selection due to the large mean distances from the cluster centroid to the cluster points. We show the robustness of our recursive reclustering technique by demonstrating that both recursive affinity propagation and recursive k-means provide a similar reduction in drive time. We validate our approach by computing exact drive times for 200 random Lowe's Home Improvement, 200 random CVS Pharmacy, and 200 random Walmart locations within the United States. We show that recursive affinity propagation shows no statistically significant difference in drives times for all three datasets. While recursive k-means shows a statistically significant difference in drive times for Lowe's, the difference in drive times is under 30 s and is insignificant to the consumer. Recursive mean shifts shows a statistically significant difference in drive times for all three datasets, however the differences are practically insignificant to the consumer and under 30 s.

The rest of this paper is organized as follows: we discuss the work related to our approach on site analysis in Sect. 2. We describe our approach on constrained recursive reclustering in Sect. 3, together with the recursive splitting of clusters, the clustering algorithms analyzed, and the constraint set used for splitting. Section 4 provides our method to evaluate the efficiency of our approach by computing drive times within a bounding box centered around a site location. In Sect. 5, we describe our dataset, and we show the computational improvements gained by performing constrained recursive reclustering of block groups. We discuss the practical and statistical differences in drive times using 200 random locations each from CVS Pharmacy, Lowe's Home Improvement, and Walmart in Sect. 6 for the original block groups, clusters obtained by traditional clustering, and clusters obtained by constrained recursive reclustering. We provide a summary and directions for future research in Sect. 7.

2   Related Work

Our work falls in the category of techniques that use census data to perform site analysis and selection. A number of approaches use census data to study the creation or propagation of food deserts. Blanchard and Lyson [4] use zip code business pattern data from the U.S. Census Bureau to determine the effect of supercenter grocery stores in reducing access to low-cost grocery stores for disadvantaged rural populations. The approach of Farber et al. [12] performs analysis of travel time from each census block in Cincinnati, Ohio to nearest supermarkets at different times of the day to investigate the occurrence of food deserts. Similarly, the approach of Jiao et al. [19] identifies food deserts by using census data to combine income with computation of travel times to supermarkets using walking, bicycling, ride transit, or driving within 10 min.

In the area of emergency care access, Branas et al. [6] use census block group data in conjunction with trauma centers and base helipads, and estimate that 69.2% and 84.1% of all U.S. residents have access to level I or II trauma centers within 45 and 60 min respectively. They estimate drive time using mathematical models that use p-norms to approximate the driving distance [25] together with drive speed estimates in urban, suburban, and rural categories. They obtain drive speed estimates by categorizing average population densities in each block group as high for urban, medium for suburban, and low for rural, and they add extra time to account for receipt of emergency call and time spent at the scene of emergency. Carr et al. [7] use emergency department (ED) location information from the National Emergency Department Inventories and the drive time estimation approach of Branas et al. [6] to determine that 71% of the U.S. population has access to an ED within 30 min, and 98% within 60 min. The approach of Nattinger et al. [27] uses the haversine formula to calculate the driving distance between hospitals offering radiotherapy services and census tracts for U.S. patients with breast cancer between 1991 and 1992. Simple distance estimation approaches using p-norm functions or haversine formula to approximate the distance between two locations do not take into account the effect of road structure and local changes in speed.

The approach of Athas et al. [2] uses the ArcGIS software to estimate shortest drive times between radiation treatment facilities and female individuals diagnosed with breast cancer in New Mexico between 1994 and 1995. While 70% of individuals were geocoded to a unique street address, the remaining 30% who had post office box or rural addresses were geocoded to the centroid of the ZIP code. Similarly, the approach of Goodman et al. [16] uses digitized road maps together with road category and traffic weightings to compute distances between the geographic centers of zip codes and nearest hospital and primary care physicians. Zip code centers are often not ideal for drive time calculations, as they may not provide optimal drive time estimates for individuals living at zip code boundaries. Instead of relying on zip code centers, our approach uses constraints on distance estimates and population counts to estimate centroids representative of block groups in urban or rural areas for more optimal drive time calculation.

The approach of Nallamothu et al. [26] obtains drive time estimates by using data on interstate, state, and local roads in Topologically Integrated Geographic Encoding and Referencing (TIGER) data from 2000 to estimate drive distances together with Census Feature Classification Codes for each road type to determine drive speeds. They use drive time estimates and population data to determine that 79% of adult population of age 18 and older lives within 60 min of a hospital that performs percutaneous coronary intervention. While their approach uses higher resolution data compared to Branas et al. [6] and Carr et al. [7] in estimating drive time, the high-resolution data introduces a higher drive time computation time, which can become intractable for repeated testing of potential new sites. Our approach resolves the issue of intractability by clustering census block groups into smaller clusters. Li et al. [23] use a map-matching algorithm to map taxicab GPS trajectories onto a road network. They use partitioning around k-medoids to iteratively solve the problem of finding the nearest locations to respond to emergency requests. While they use the k-medoids clustering algorithm [28], their clustering is not meant to reduce the quantity of data used in drive time computation as in our approach. Instead, they use the update step of k-medoids clustering to iteratively converge to the best response locations for emergency requests.

While our work focuses on using census data and drive time for site analysis, several approaches on site selection integrate a variety of external features in determining the quality of a site. Xu et al. [40] use features such as distances to the city center, traffic, POI density, category popularity, competition, area popularity, and local real estate pricing to determine the feasibility of a location. The approach of Karamshuk et al. [21] uses features mined from FourSquare along with supervised learning approaches to determine the optimal location of a retail store. A number of approaches use information obtained from social media platforms, such as popularity based on user reviews [37] or based on number of user check-ins and viability of location as obtained from Twitter and FourSquare [10,21,30,38,43,44]. Given the large number of features that may be used to evaluate a site, some approaches use fuzzy techniques to determine sites that show the best compromise between various site selection criteria [20]. Approaches have used fuzzy techniques to determine the optimal number of fire stations at an airport [34], and optimal locations of new convenience stores [22] and factories [8,42]. There also exist approaches that use user expertise to weight location selection criteria in reaching a compromise on best location [1,35,41].

While our approach is designed to provide drive time computations for user-provided site queries, there exist approaches that perform automated determination of an optimal site query given a set of existing sites and current customer locations [14,39]. The approach of Ghaemi et al. [15] performs optimal query estimation while addressing issues caused by movement of existing sites and customers. Banaei et al. [3] perform reverse skyline queries to incorporate additional criteria such as distance to location and distance to competitors in optimal location query estimation.

3   Constrained Recursive Reclustering

Our approach on recursive reclustering uses a user-provided constraint set to recursively split an initial set of clusters into a final set that satisfies the constraint set. Our approach is flexible enough to accommodate any form of clustering algorithm, and can use constraint sets defined using operations such as intersections, unions, and complements on a number of inequality constraints. We discuss the recursive cluster splitting approach in Subsect. 3.1. The clustering algorithms analyzed in this work are discussed in Subsect. 3.2, while the particular constraint set used in this work is discussed in Subsect. 3.3. Unlike existing approaches on constrained clustering that address the issue of satisfying user-defined constraints and user-provided parameters such as number of clusters [5,36], our approach focuses on addressing user-defined constraints for non-parametric clustering, i.e., in our approach, the user does not need to provide parameters such as the number of k-means clusters or the mean shift kernel bandwidth.

3.1   Recursive Cluster Splitting

For any particular clustering algorithm discussed in Subsect. 3.2, our reclustering approach recursively performs cluster splitting approach starting from an initial set of clusters generated using the approach discussed in Subsect. 3.2 for each algorithm. In addition to the initial clusters, our approach uses a constraint set Ω that can be defined for each cluster using operations such as intersections, unions, and/or complements on a number of inequality constraints. Starting from the initial cluster set, our approach splits each cluster into a set of child clusters using the particular clustering algorithm if the constraint set Ω is not met. The clustering approach is performed recursively till Ω is met for each resultant cluster.

Algorithm 1 summarizes the steps of our recursive clustering approach. The algorithm adapts the recursive clustering splitting algorithm from our prior work [18] to work with any clustering algorithms and constraint sets. The initial clustering algorithm runs in $O(tn^2)$ time and produces k clusters, where t represents the number of iterations until convergence and n represents the number of samples. Each of the k clusters is reclustered in $O(tm_i^2)$ time, where m_i represents the number of points in the ith cluster and $i \in [1, 2, \cdots, k]$.

While our recursive reclustering approach is related to divisive hierarchical clustering techniques [31], it differs from them in that hierarchical clustering approaches focus on generating clusters that meet distance constraints intrinsic to the cluster points, while our approach integrates external user-defined constraints. As discussed in Sect. 5, we compare the effect of our recursive reclustering approach on drive time computation to using the original block groups, and to using traditional clustering as obtained by running the clustering algorithms in Subsect. 3.2 without reclustering.

3.2   Clustering Algorithms Analyzed in This Work

In this paper, we compare the performance of three different clustering algorithms—affinity propagation [13], k-means [24], and mean shift [11].

Affinity Propagation. The approach of affinity propagation, proposed by Frey and Dueck [13], uses message passing between data points to select a set of exemplar points. Each exemplar point is representative of a cluster of data points in its vicinity. The advantage of the method is that it does not require user-specified parameters such as the number of clusters as in k-means or the kernel bandwidth as in mean shift. The method iteratively refines estimates on exemplar points by updating responsibility messages that data points send to candidate exemplar points on the suitability of the exemplar points to represent the data points, and availability messages that candidate exemplar points send back to data points to reflect how well the candidate exemplar represents each data point based on accumulated responsibility evidence from other data points. In this work, we use the affinity propagation function implemented in the scikit-learn toolbox [29], with a damping factor of 0.9 to reduce numerical oscillations in updates of the responsibility and availability. We use 2000 maximum iterations, and 200 convergence iterations, i.e., iterations over which the number of clusters remain consistent for convergence.

The affinity propagation algorithm does not directly accommodate external user-defined constraints. We find that in our work, the clusters generated by affinity propagation are large and do not satisfy the constraint set discussed in Subsect. 3.3. We use the recursive clustering approach discussed in Subsect. 3.2 to perform further clustering using affinity propagation till the constraint set is met.

k-means. We use the k-means function in the scikit-learn toolbox that implements Lloyd's algorithm [24] to partition the region in Euclidean space containing the locations of the points into a user-specified number of clusters, k. Given an initialization for the cluster centroids, Lloyd's algorithm iteratively computes the Voronoi diagram for the k clusters, and updates the centroid given each Voronoi cell. The computation of the Voronoi diagram effectively assigns each sample to its nearest centroid. The algorithm proceeds till the k-means objective function defined as the within cluster sum-squared Euclidean distance converges, i.e., till the difference between consecutive values of the function falls below a tolerance level. Since changes in the initial centroid locations yield different values of the objective function, we use m initializations of the k-means algorithm, and select the initialization that yields the lowest value of the k-means objective function. In this work, we use the default value of $m = 10$ in the scikit-learn toolbox, together with the default values of 0.001 for tolerance level and 300 for maximum number of iterations.

The principal challenge of the k-means approach is that the number of clusters k needs to be pre-specified *a priori*. In this work, we use the approach of silhouettes [32] to select the best number of clusters between 2 and an upper bound K on the number of clusters. For a candidate number of clusters k, the silhouette score is obtained as the difference between the mean nearest-cluster Euclidean distance b and the mean intra-cluster Euclidean distance a, scaled by the maximum value of a and b to obtain a value between -1 and 1. Values near 1 indicate good clusters, values near 0 indicate overlapping clusters, while values near -1 indicate that a point is incorrectly assigned to a cluster. We conduct k-means for $k \in [2, K]$, and select the best value of number of clusters $k^\star$ as the k for which the silhouette score is highest.

While one choice for the value of K is the total number of data points n, running k-means and silhouette score calculation for $k \in [2, n]$ is computationally intensive. We perform silhouettes-based k-means clustering using $k \in [2, n]$ for a random sampling of 25% U.S. states. We find that all U.S. states in the random sampling have a value of $k^\star \leq 60$, while 70% of the U.S. states in the random sampling have a value of $k^\star \leq 20$. The constraint set Ω discussed in Subsect. 3.3 is not met for all clusters in any of the U.S. states, indicating that recursive clustering is necessary with silhouette-based k-means. To optimize between run-time and optimal selection of number of clusters, we set K to 20 for all 50 U.S. states in the initial and recursive clustering steps.

Mean Shift. The mean shift approach for clustering [11] is a mode-seeking algorithm that recursively updates the means of a collection of shifted kernel functions in representing the maxima or modes or a density function. Each cluster provided as the result of the mean shift clustering algorithm is given as the set of data points closest to the mean of the kernel function representing that cluster. In this work, we use radial basis function (RBF) kernels to perform mean shift as implemented in the `scikit-learn` toolbox. While the mean shift approach is considered a non-parametric technique in that it does not require specification of the number of clusters, it does require specifying the bandwidth of the kernel used in mode-seeking. Higher bandwidth kernels yield larger clusters, while lower bandwidth kernels yield smaller clusters. We use the bandwidth estimation function built into `scikit-learn` that uses the 30th percentile of pairwise distances as the band-width.

3.3   Constraint Set Used in This Work

The constraint set Ω in this work is defined as the union between a cluster points count constraint ω_c and the intersection of a population count constraint ω_p and a distance constraint ω_p, i.e.,

$$\Omega = \omega_c \cup (\omega_d \cap \omega_p). \tag{1}$$

This induces our approach to perform recursive clustering till either the number of points in the cluster falls below a user-provided threshold, i.e., till the cluster points count constraint ω_c is satisfied, or till the total population and the mean haversine distance between the cluster points to the cluster centroid both fall below user-provided thresholds, i.e., till both ω_p and ω_d are satisfied.

Algorithm 1. Recursive Cluster Splitting.

Input: Sets of latitudes and longitudes for initial cluster points
$$\{\{(\phi_i, \lambda_i) : i \in \mathcal{I}_{c_{init}}\} : c_{init} \in \mathcal{C}_{init}\},$$
Set of latitudes and longitudes for initial cluster centroids
$$\{(\phi_{c_{init}}, \lambda_{c_{init}} : c_{init} \in \mathcal{C}_{init}\},$$
and user-provided bounds c_{bound}, d_{bound}, and p_{bound}
Output: Set of final clusters, $\mathcal{O}$

1   **for** $c_{init} \in \mathcal{C}_{init}$ **do**
2      $\mathcal{P}_{c_{init}} \leftarrow \{(\phi_i, \lambda_i) : i \in \mathcal{I}_{c_{init}}\}$
3      $\mathcal{O} = \text{split}(\mathcal{P}_{c_{init}}, O)$
4   **return** $\mathcal{O}$
   end

   Procedure split($\mathcal{P}_c$, $\mathcal{O}$)
1   Compute the constraint set Ω using Equation (1)
2   **if** *the constraint set Ω is not met* **then**
3   Split cluster represented by points in $\mathcal{P}_c$ by clustering them into smaller clusters $\{P_{\bar{c}} : \bar{c} \in \mathcal{C}_c\}$ using the clustering algorithm
4   **for** $\bar{c} \in \mathcal{C}_c$ **do**
5   **return** split($\mathcal{P}_{\bar{c}}, \mathcal{O}$)
         end
     else
6          $\mathcal{O} \leftarrow \mathcal{O} \cup \mathcal{P}_c$
7   **return** $\mathcal{O}$
     end

The cluster points count constraint, ω_c for a particular cluster c is defined as

$$\omega_c : |\mathcal{I}_c| \leq c_{bound}, \tag{2}$$

where $\mathcal{I}_c$ represents the index set to the number of points in the cluster c, $|\mathcal{I}_c|$ represents the size of $\mathcal{I}_c$ or the count of the number of points, and c_{bound} represents the user-provided upper bound on the cluster count.

The population count constraint, ω_p for cluster c is defined as

$$\omega_p : p_c \leq p_{bound}, \tag{3}$$

where p_c represents the user population in that cluster as obtained by summing the population counts of all block groups in cluster c, while p_{bound} represents the user-provided upper bound on the population count.

The distance constraint, ω_d for cluster c is defined as

$$\omega_d : \frac{1}{|\mathcal{I}_c|} \sum_{i \in \mathcal{I}_c} \left(2R \operatorname{atan2} \left(\sqrt{a_i}, \sqrt{1 - a_i} \right) \right) \leq d_{bound}, \tag{4}$$

where the term on the left hand side of the inequality represents the mean haversine distance between the cluster centroid and the cluster points indexed by the set $\mathcal{I}_c$, and d_{bound} represents the user-provided upper bound on the mean haversine distance. In Inequality (4), a_i represents the haversine of the central angle between each point represented by its latitude ϕ_i and longitude λ_i to its cluster centroid represented by ϕ_c and λ_c, and is computed as

$$a_i = \sin^2 \frac{\phi_c - \phi_i}{2} + \cos \phi_i \cdot \cos \phi_c \cdot \sin^2 \frac{\lambda_c - \lambda_i}{2}. \tag{5}$$

For all results in this work, the distance d_{bound} is set to 5 miles, the population count bound p_{bound} is set to $20,000$, and the cluster points count bound c_{bound} is set to 10 points.

4 Drive Time Computation

We evaluate our approach by performing drive time computation from site locations for a particular business to all points within a bounding box of a user-specified half-size centered at the business location. The points refer to either the original block groups or the cluster centroids obtained using recursive reclustering on the clustering algorithms discussed in Subsect. 3.2, and represent customers most likely to visit the business location. Given a bounding box half-size d, and the location of the business represented by latitude ϕ_1 and longitude λ_1, we use the inverse haversine formula to obtain the north-east and south-west locations of the bounding box, denoted by latitudes ϕ_{max} and ϕ_{min} and longitudes λ_{max} and λ_{min} respectively, as described in Algorithm 2 and as discussed in our prior work [18]. We set the bounding box half-size d to 5 miles. As a note, while we use the haversine distance formula as a heuristic to compute the bounding box and to split the clusters as part of the constraint set in Sect. 3.3, we do not use the haversine distance between the starting and ending points to compute the drive times as in the approach of Nattinger et al. [27], since the start-to-end haversine distance does not account for the road structure and changing speed limit between the two locations. Instead, we use the Google Maps Distance Matrix API [17] to generate drive times from the business location to all points within the bounding box. Algorithm 1 generates the bounding box within $O(1)$ time, while the drive time computations using Google Maps API are performed in $O(l)$ time, where l is the number of points in the bounding box. Our approach of using clustering accelerates the drive time calculations in comparison to the original block groups due to the $O(l)$ dependence of the drive time queries.

5 Traditional vs. Recursive Reclustering Results

We compare traditional clustering to recursive reclustering to determine how well each approach works in reducing the number of census block groups from 220,334 to an optimized set. The optimal number of clusters is one where the drive time computed using the census dataset and the clustered dataset shows no practical or statistically significant difference. Traditional clustering is done as a single pass using affinity propagation, k-means, and mean shift. Our recursive reclustering approach takes the sub-optimal clusters generated by the traditional clustering approach and continues to recluster them until the user specified stopping constraints are satisfied. The user specified constraints are described in Sect. 3.

Algorithm 2. Bounding Box Computation as performed in Gwinn et al. [18].

Parameters: $MINLAT$ (min latitude):$-90°$,
$MAXLAT$ (max latitude):$90°$,
$MINLON$ (min longitude):$-180°$,
$MAXLON$ (max longitude):$180°$,
R (radius of earth): $6,371$ km.

Input: Distance d and location (ϕ_1, λ_1)

1 $\phi = \frac{d}{R}$
2 $\phi_{\min} = \phi_1 - \phi$
3 $\phi_{\max} = \phi_1 + \phi$
4 **if** $\phi_{min} > MINLAT \wedge \phi_{max} < MAXLAT$ **then**
5 $\quad$ $\lambda = \sin^{-1}\left(\frac{\sin\phi}{\cos\phi_1}\right)$
6 $\quad$ $\lambda_{\min} \leftarrow \lambda_1 - \lambda$
7 $\quad$ **if** $\lambda_{min} < MINLON$ **then**
8 $\quad\quad$ | $\lambda_{\min} \leftarrow \lambda_{\min} + 2\pi$
$\quad$ **end**
9 $\quad$ $\lambda_{\max} \leftarrow \lambda_1 + \lambda$
10 $\quad$ **if** $\lambda_{max} > MAXLON$ **then**
11 $\quad\quad$ | $\lambda_{\max} \leftarrow \lambda_{\max} - 2\pi$
$\quad$ **end**
$\quad$ **else**
12 $\quad$ $\phi_{\min} \leftarrow \max(\phi_{\min}, \text{MINLAT})$
13 $\quad$ $\phi_{\max} \leftarrow \min(\phi_{\max}, \text{MAXLAT})$
14 $\quad$ $\lambda_{\min} \leftarrow \text{MINLON}$
15 $\quad$ $\lambda_{\max} \leftarrow \text{MAXLON}$
$\quad$ **end**

5.1 Traditional Clustering Results

As shown in Table 1, affinity propagation reduces the census block groups from 220,334 block groups to 2,113 clusters. In Table 2, we show that affinity propagation generates a larger number of clusters in states with a larger number of block groups. We find a strong positive correlation of 0.73 between the number of census block groups and affinity clusters. The clusters generated by the traditional affinity propagation algorithm are sub-optimal, with large distances between cluster centroids and cluster points. For example, as shown in Table 3 for a sparsely populated state, such as Alaska, the average cluster centroid to cluster points is 57.0 miles with a maximum distance of 157.9 miles. We observe large distances for densely populated states, such as Rhode Island, with an average centroid to cluster point distance of 3.3 miles and a maximum distance of 6.8 miles. When considering the average across all states, the average maximum distance from the cluster centroid is 41.1 miles, and the average cluster centroid to cluster points distance is 15.9 miles.

Table 1. Number of clusters generated using traditional clustering and our recursive reclustering approach. Traditional clustering is run in a single pass, while recursive reclustering continues reclustering each cluster until the user specified constraints are satisfied.

	Census block group	Affinity propagation	k-Means	Mean shift
Traditional	220,334	2,113	290	43,852
Recursive reclustering	N/A	41,442	36,666	197,675

Table 2. Block groups (BG), initial cluster counts using traditional clustering, and final cluster counts using reclustering with affinity propagation (AP), k-means, and mean shift (MSh) on 50 U.S. states, District of Columbia (DC), and Puerto Rico (PR).

State	AL	AK	AZ	AR	CA	CO	CT	DE	DC	FL	GA	HI	ID
BG	3438	534	4178	2147	23212	3532	2585	574	450	11442	5533	875	963
AP Initial	39	21	53	35	70	64	37	12	17	50	57	14	15
AP Final	704	115	872	448	4090	736	422	112	80	2150	1167	167	227
k-Means Initial	9	3	5	13	2	2	2	2	2	7	2	5	4
k-Means Final	588	114	651	403	3964	560	390	91	61	1910	955	152	199
MSh Initial	965	45	154	645	4039	659	1222	38	221	1470	1545	8	13
MSh Final	3061	254	3565	1875	22122	3039	2469	415	416	10567	5245	609	582
State	IL	IN	IA	KS	KY	LA	ME	MD	MA	MI	MN	MS	MO
BG	9691	4814	2630	2351	3285	3471	1086	3926	4985	8205	4111	2164	4506
AP Initial	72	46	38	34	40	34	27	57	54	72	67	32	40
AP Final	1809	919	555	503	644	633	222	722	848	1471	829	459	881
k-Means Initial	2	12	16	3	10	4	2	3	2	4	3	10	5
k-Means Final	1550	825	464	436	554	561	216	655	725	1259	723	421	747
MSh Initial	1770	1399	612	357	1337	147	375	1038	972	1313	1124	606	722
MSh Final	8897	4425	2098	1720	2813	2706	933	3628	4862	7613	3744	1742	3862
State	MT	NE	NV	NH	NJ	NM	NY	NC	ND	OH	OK	OR	PA
BG	842	1633	1836	922	6320	1449	15464	6155	572	9238	2965	2634	9740
AP Initial	18	34	65	24	61	20	68	57	14	55	44	37	58
AP Final	203	374	279	176	1035	307	2541	1198	164	1662	637	581	1702
k-Means Initial	13	2	3	6	2	15	2	2	18	4	20	2	2
k-Means Final	168	298	291	170	934	277	2289	1076	134	1376	518	460	1566
MSh Initial	60	300	11	368	1488	91	1872	3308	132	1795	839	413	2173
MSh Final	464	1359	1569	782	5934	972	14525	6041	365	8590	2604	2192	9008
State	PR	RI	SC	SD	TN	TX	UT	VT	VA	WA	WV	WI	WY
BG	2594	815	3059	654	4125	15811	1690	522	5332	4783	1592	4489	410
AP Initial	37	27	34	15	26	76	51	16	41	47	24	54	13
AP Final	467	117	621	177	805	2994	331	108	1028	919	280	852	99
k-Means Initial	2	2	13	2	3	6	2	2	6	2	3	2	19
k-Means Final	437	98	538	133	711	2804	331	95	935	780	263	732	77
MSh Initial	635	177	520	131	329	3098	178	86	1550	657	94	721	30
MSh Final	2223	710	2514	449	3689	14581	1313	336	4937	4312	1170	3565	209

Table 3. Maximum and mean values of distance between the centroid and the cluster points for initial clusters using traditional clustering and final clusters using reclustering with affinity propagation (AP), k-means, and mean shift (MSh) on 50 U.S. states, District of Columbia (DC), and Puerto Rico (PR).

State	AL	AK	AZ	AR	CA	CO	CT	DE	DC	FL	GA	HI	ID
AP Initial Max	39.83	157.86	39.96	42.19	52.18	33.37	12.91	14.79	1.96	42.37	35.71	26.22	71.56
AP Initial Mean	17.33	57.02	15.27	18.29	16.20	11.53	5.71	6.06	0.98	15.06	14.81	11.02	23.82
AP Final Max	5.17	27.14	2.86	6.18	1.73	3.00	2.55	2.57	0.67	2.38	3.89	2.82	5.25
AP Final Mean	2.77	11.37	1.47	3.14	0.90	1.47	1.40	1.42	0.38	1.31	2.07	1.42	2.48
k Initial Max	96.62	842.15	196.60	76.32	432.38	282.96	74.63	53.34	6.95	161.22	251.95	60.81	198.29
k Initial Mean	37.14	250.58	52.56	31.49	88.40	86.85	27.50	20.08	3.17	49.26	93.10	20.90	57.31
k Final Max	5.47	20.22	3.65	5.99	1.64	3.90	2.48	2.92	0.74	2.47	4.35	2.75	5.59
k Final Mean	3.51	13.16	2.19	3.74	1.02	2.28	1.56	1.75	0.46	1.56	2.74	1.66	3.40
MSh Initial Max	0.30	30.99	6.49	0.45	0.13	1.09	0.04	2.27	0.02	0.34	0.16	32.90	44.39
MSh Initial Mean	0.13	15.00	3.14	0.19	0.06	0.57	0.02	0.94	0.01	0.15	0.08	12.36	17.87
MSh Final Max	0.10	4.94	0.19	0.14	0.02	0.15	0.01	0.16	0.01	0.03	0.04	0.29	0.65
MSh Final Mean	0.06	2.56	0.12	0.09	0.01	0.09	0.01	0.10	0.00	0.02	0.03	0.19	0.39

State	IL	IN	IA	KS	KY	LA	ME	MD	MA	MI	MN	MS	MO
AP Initial Max	31.45	32.44	44.08	53.30	36.65	45.02	34.03	15.46	14.32	37.99	38.22	44.56	49.22
AP Initial Mean	13.00	13.94	19.54	22.40	16.40	16.20	15.48	6.62	5.98	14.36	16.48	18.69	20.86
AP Final Max	2.82	3.87	5.88	5.47	5.01	4.38	6.99	2.37	2.01	3.62	5.10	6.26	4.75
AP Final Mean	1.41	1.98	2.83	2.51	2.72	2.26	3.70	1.28	1.08	1.90	2.56	3.31	2.47
k Initial Max	254.97	69.59	71.06	210.41	76.26	133.40	208.90	113.30	115.57	210.21	256.04	77.09	162.90
k Initial Mean	84.21	27.39	30.71	84.97	33.39	52.14	73.44	40.78	36.48	72.79	82.67	36.70	55.42
k Final Max	2.80	3.78	6.31	5.33	5.29	4.41	6.12	2.28	2.14	3.90	5.14	5.96	5.17
k Final Mean	1.73	2.37	3.81	3.32	3.40	2.74	3.99	1.44	1.29	2.42	3.28	3.82	3.21
MSh Initial Max	0.20	0.18	0.54	1.20	0.11	2.54	0.51	0.13	0.11	0.33	0.30	0.49	0.64
MSh Initial Mean	0.08	0.08	0.30	0.53	0.05	1.30	0.23	0.06	0.06	0.14	0.12	0.27	0.24
MSh Final Max	0.06	0.05	0.27	0.33	0.10	0.21	0.12	0.03	0.01	0.04	0.10	0.30	0.16
MSh Final Mean	0.04	0.03	0.18	0.20	0.06	0.14	0.08	0.02	0.01	0.03	0.06	0.20	0.10

State	MT	NE	NV	NH	NJ	NM	NY	NC	ND	OH	OK	OR	PA
AP Initial Max	97.59	47.43	9.48	21.25	13.25	75.18	33.43	35.65	83.41	33.36	41.41	44.32	33.32
AP Initial Mean	33.70	19.24	3.77	9.22	5.52	22.55	12.24	14.45	30.55	13.51	16.16	13.69	12.97
AP Final Max	8.63	5.39	2.64	4.58	1.73	4.29	2.17	4.28	8.61	3.01	4.69	3.11	3.08
AP Final Mean	3.74	2.63	1.27	2.45	0.93	2.11	1.14	2.35	4.22	1.59	2.32	1.49	1.61
k Initial Max	116.31	264.96	262.96	54.46	91.25	95.17	389.41	280.25	68.91	148.54	65.81	335.22	220.78
k Initial Mean	39.30	108.05	51.87	22.25	33.46	30.02	84.08	89.25	27.05	49.94	27.98	113.14	74.56
k Final Max	9.79	6.29	3.37	4.12	1.70	4.57	2.15	4.26	9.75	3.32	5.14	4.15	2.93
k Final Mean	5.87	3.98	2.16	2.58	1.04	2.84	1.34	2.75	6.80	2.05	3.22	2.42	1.84
MSh Initial Max	11.84	1.57	65.72	0.30	0.07	6.46	0.22	0.04	2.59	0.15	0.35	1.80	0.12
MSh Initial Mean	5.65	0.79	25.43	0.15	0.03	2.96	0.08	0.02	1.25	0.07	0.12	0.97	0.04
MSh Final Max	1.60	0.33	0.13	0.12	0.01	0.46	0.03	0.01	1.38	0.02	0.12	0.20	0.03
MSh Final Mean	0.97	0.20	0.08	0.07	0.00	0.28	0.02	0.00	0.86	0.01	0.08	0.13	0.02

State	PR	RI	SC	SD	TN	TX	UT	VT	VA	WA	WV	WI	WY
AP Initial Max	11.66	6.83	35.47	76.99	49.62	65.64	22.56	27.08	38.32	39.51	37.38	39.34	71.77
AP Initial Mean	5.09	3.27	14.00	30.61	20.33	23.47	8.76	12.86	15.09	14.78	16.41	15.89	23.71
AP Final Max	1.95	2.29	4.42	8.50	4.46	3.22	3.53	7.14	3.63	3.04	6.63	4.68	5.98
AP Final Mean	1.03	1.26	2.39	4.26	2.40	1.61	1.73	3.72	1.96	1.55	3.46	2.41	2.37
k Initial Max	35.34	58.78	263.30	169.23	298.00	311.28	94.10	127.22	211.39	135.02	274.62	48.80	95.73
k Initial Mean	13.28	22.93	105.88	70.18	86.66	72.75	42.12	39.29	90.46	53.16	96.11	17.97	25.25
k Final Max	2.41	4.46	10.50	4.45	3.14	2.96	6.73	3.55	3.35	6.13	4.88	8.42	1.82
k Final Mean	1.45	2.80	6.70	2.86	1.96	1.94	4.29	2.27	2.06	3.95	3.06	4.50	1.10
MSh Initial Max	0.29	0.47	2.99	1.47	0.27	2.97	1.50	0.20	0.64	3.46	0.57	17.78	0.11
MSh Initial Mean	0.15	0.28	0.95	0.62	0.12	1.73	0.84	0.10	0.21	1.59	0.28	11.30	0.06
MSh Final Max	0.07	0.16	1.08	0.08	0.05	0.31	0.59	0.06	0.10	0.27	0.19	1.13	0.04
MSh Final Mean	0.05	0.11	0.70	0.05	0.03	0.20	0.37	0.04	0.06	0.18	0.12	0.64	0.03

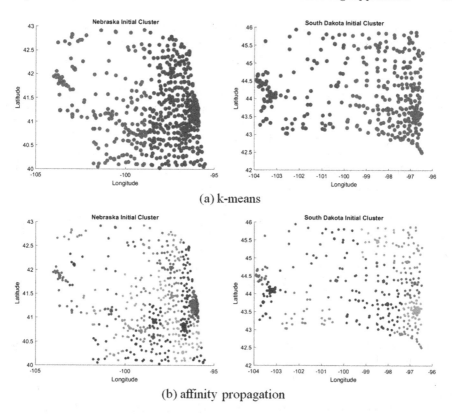

(a) k-means

(b) affinity propagation

Fig. 2. Comparison of clustered census block groups for South Dakota and Nebraska. Both states have a land area of 77,000 miles, but South Dakota has a population of 869,666 while Nebraska has a population of 1,920,076. (a) k-means generates 2 clusters for both states and is unable to factor in the distribution of population. (b) Affinity propagation generates 34 clusters for the more densely populated state of Nebraska and 15 clusters for South Dakota. [Figure best viewed in color]. (Color figure online)

From Table 1, we see that k-means reduces the census block groups 290 clusters. From Table 2, we see that k-means generates between 2 and 20 clusters with a median of 3 clusters. The clusters generated by the traditional k-means algorithm are highly sub-optimal, with an average distance of 58 miles between the cluster centroid and cluster points. As shown in Table 3, for Alaska we find the average cluster centroid to cluster points distance is 250.6 miles with a maximum distance of 842.1 miles. Even in densely populated states, such as Rhode Island, we find a large average cluster centroid to cluster points distance of 22.9 miles and a maximum of 58.8 miles. When considering the average across all states, the average maximum distance from the cluster centroid is 177.1 miles, and the average cluster centroid to cluster points distance is 58.0 miles.

As shown in Table 1, mean shift reduces the census block groups to 43,852 clusters. In Table 2, we observe that mean shift generates between 8 and 4,039 clusters, with a strong positive correlation of 0.87 between census block group count and cluster count. One of the challenges of mean shift is the generation of a large number of single point clusters due to the points being located in areas of low density.

We find that affinity propagation is the ideal initial clustering algorithm, as unlike k-means the user does not need to specify the number of clusters, and unlike mean shift it does not generate a large number of clusters of size 1. As shown in Fig. 2, affinity propagation is more adaptive to population distribution within a state. South Dakota and Nebraska both have a land area close to 77,000 square miles, however Nebraska is a more densely populated state with over 2× the population of South Dakota. While k-means generates 2 clusters for both states, affinity propagation generates 34 clusters for Nebraska, the more densely populated state, and 15 clusters for South Dakota, the more sparsely populated state. We further illustrate this in Fig. 3, where we show the differences in clustering using affinity propagation, k-means and mean shift in densely populated states such as Rhode Island, California, and Illinois, and sparsely populated states such as Nevada, North Dakota, and Wyoming.

5.2   Recursive Clustering Results

As shown in Table 1, our recursive reclustering approach using affinity propagation reduces the census dataset from 220,334 block groups to 41,442 optimized clusters by reclustering the initial 2,113 clusters. Our approach provides a 81.2% reduction in the size of the dataset, with the highest reduction of 85.6% in Rhode Island where the 815 block groups are reduced to 117 clusters. The lowest reduction in the number of census block groups is in North Dakota, with 572 block groups being reduced to 164 clusters, i.e. a reduction of 71.3%. As shown in Table 3, the average maximum distance from the cluster centroid for all states is 4.6 miles, and the average cluster centroid to cluster points distance is 2.3 miles. For a sparsely populated state, such as Alaska the average distance from the cluster centroid to the cluster points is 11.4 miles with an average maximum of 27.1 miles. In a densely populated state, such as the District of Columbia, the average maximum distance from the cluster centroid is 0.7 miles with the average distance from the cluster centroid to all points being 0.4 miles. In densely populated states, such as the District of Columbia, small distances can have longer drive times, hence a low cluster centroid to cluster point distance is ideal.

In Table 1, we show that our recursive reclustering approach using k-means reclusters the 290 sub-optimal clusters into 36,666 optimal clusters. This results in a 83.4% reduction in the census dataset when compared to the 220,334 census block groups. The highest reduction is obtained in Rhode Island, with our approach reducing the 815 original block groups to 98 clusters, i.e. an 88.0% reduction. We see the lowest reduction in North Dakota, with the original 572 block groups being reduced to 134 clusters, i.e. a reduction of 76.6%. As shown in Table 3, the average maximum distance from the cluster centroid for all states is 4.7 miles, and the average cluster centroid to cluster points distance is 3.0 miles. Similar to our recursively reclustered affinity propagation results, we see the

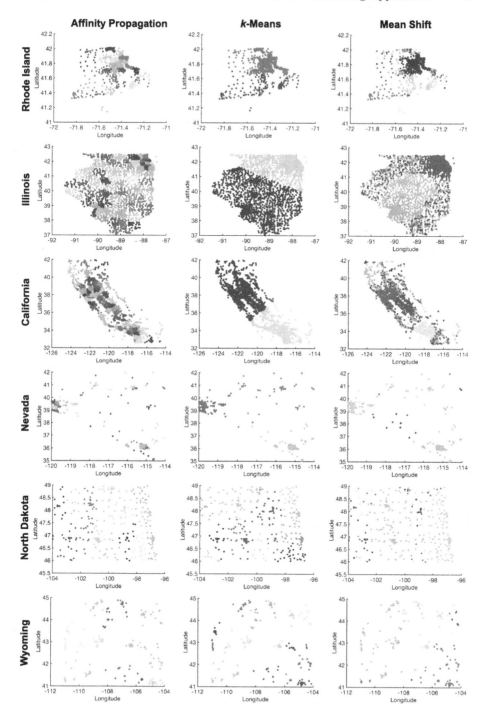

Fig. 3. Data clustered without recursion using affinity propagation, k-means, and mean shift. [Figure best viewed in color]. (Color figure online)

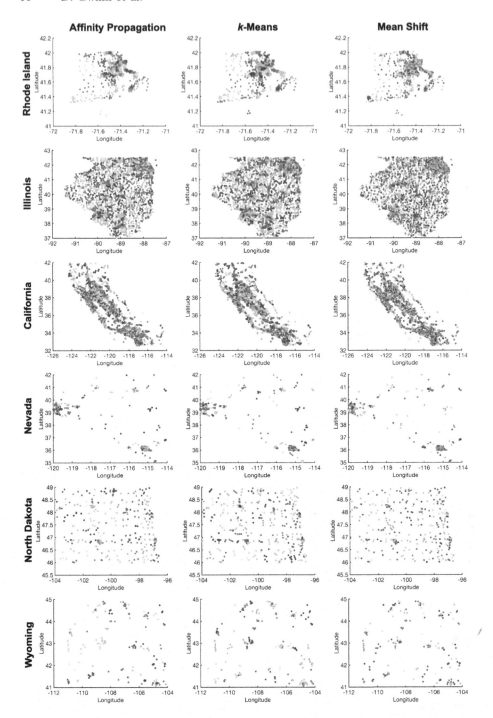

Fig. 4. Data clustered by recursive reclustering using affinity propagation, k-means, and mean shift. [Figure best viewed in color]. (Color figure online)

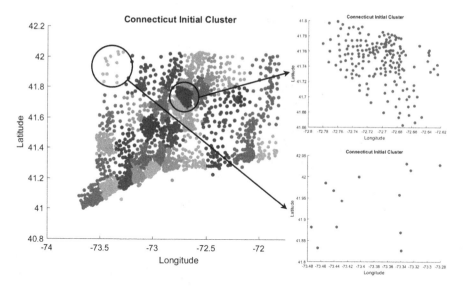

Fig. 5. A densely populated area contains several block groups in close proximity, while a sparsely populated area has larger distances between block groups. In our approach the densely populated area shown in the top right is reclustered further into smaller clusters to ensure each cluster point is less than 5 miles from the cluster centroid and the total population of the cluster is below 20,000. The sparsely populated area shown on the bottom right will also be reclustered using our approach, however our algorithm generates fewer sub clusters. [Figure best viewed in color] [18]. (Color figure online)

highest average maximum distance of 20.2 miles and highest average cluster centroid to cluster points distance of 13.2 miles in Alaska. We see the lowest average maximum distance of 0.7 miles and lowest average cluster centroid to cluster points distance of 0.4 miles in the District of Columbia. We find a strong positive correlation of 0.997 for the number of clusters generated by our recursively reclustered affinity propagation and k-means algorithms. While k-means generates the fewest number of clusters, it requires the user to provide the initial number of clusters and as a result we find affinity propagation to be the better algorithm.

From Table 1, we show that our recursive reclustering approach using mean shift reclusters the 43,582 sub-optimal clusters into 197,675 clusters which results in a 10.2% reduction in the census block group dataset. As shown in Figs. 3 and 4, our approach does not recombine single point clusters, as a result single point clusters generated by the traditional mean shift algorithm is left as-is. From Table 3, we note that recursively reclustered mean shift shows the lowest distances for both the average cluster centroid to cluster points and maximum cluster centroid to cluster point distance. However, this distance is biased by the presence of a large number of single point clusters where the cluster centroid to cluster point distance and maximum cluster centroid to cluster point distance is 0.

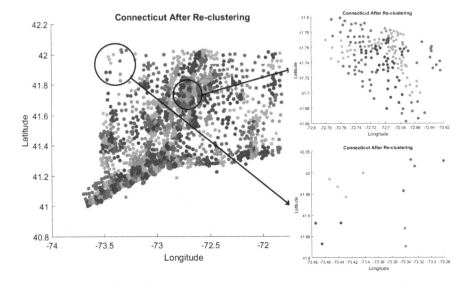

Fig. 6. A densely populated area contains several block groups in close proximity, while a sparsely populated area has larger distances between block groups. By reclustering the densely populated area into multiple smaller clusters, we ensure that the drive time differences between the raw census data and clustered data are minimized. [Figure best viewed in color] [18]. (Color figure online)

5.3   Differences in Block Group Reduction

For all three recursive reclustering approaches, the variances in census block group reduction are due to the differences in land area and population distribution. From Table 2, we note that the highest percentage reduction in block groups is in Rhode Island. As shown in Figs. 3 and 4, Rhode Island is a densely populated state with 1,021 individuals per square mile. Thus, fewer clustered block groups can be used to represent the state. States with low population density, such as Wyoming with 6 individuals per square mile show the lowest reduction. Finally, states such as Nevada where the population is concentrated in a few regions, we also observe a larger percentage reduction in the number of census block groups.

In Figs. 5 and 6, we illustrate how our recursive reclustering approach handles localities with different population densities in the state of Connecticut using affinity propagation. We show a densely populated area, Hartford, and a sparsely populated area, Salisbury, in Figs. 5 and 6. As shown in Fig. 5, a densely populated area has numerous block groups in close proximity, while a sparsely populated area has larger distances between block groups. Thus, as shown in Fig. 6, our approach reclusters the densely populated area into a higher number of clusters and the sparsely populated into a smaller number of clusters based on the constraints described in Sect. 3.

Fig. 7. Geographic positions of all locations (blue) and 200 random locations chosen (black) for CVS Pharmacy, Lowe's Home Improvement, and Walmart. The sampled random locations represent the weighting of the original locations across the U.S. (Color figure online)

6 Evaluation

In order to evaluate the effectiveness of our approach, we sample 200 random locations each from three major retailers in the United States—CVS Pharmacy, Lowe's Home Improvement, and Walmart—which each have 9,807, 1,741, and 5,746 locations respectively. We obtain a total of 600 locations. As shown in Fig. 7, the random locations represent the weighting of the original distribution of locations across the country.

A typical location selection procedure evaluates the effectiveness of a location by placing a trade area at a given radius around a location and computing the average drive time to the intended location for each potential customer within the trade area. We use the Google Maps Distance Matrix API [17] to compute the drive times for the census block group, traditional clustered, and recursively reclustered data to existing business locations using the approach discussed in Sect. 4. Since our approach in Sect. 4 yields a bounding box as opposed to a circle, we use a half-size for the bounding box of 5 miles instead of a radius.

6.1 Comparison to Traditional Clustering

We use a paired t-test to determine if the differences in drive times for census block group and traditionally clustered data are significantly different by testing the following hypotheses:

NULL: the mean drive time for census block group data is no different from the mean drive time for traditionally clustered data.

Alternate: the mean drive time for census block group data is different from the mean drive time for traditionally clustered data.

From Table 4, we observe that all CVS, Lowe's, and Walmart locations contain at least one census block group within 5 miles. We note that traditional clustering algorithms are ill suited as over 40% of the locations no longer have a clustered block group within 5 miles when using affinity propagation, 90% when using k-means, and over 70% when using mean shift.

From Table 5, we see no statistically significant differences in drive time for all three datasets using all three algorithms. However, 91 CVS, 93 Lowe's, and

Table 4. Comparison of the number of actual retail store locations, and the number of retail stores with at least one census block group or clustered block group within 5 miles.

Dataset	Actual	Census	Traditional			Recursively reclustered		
			AP	k-Means	Mean Shift	AP	k-Means	Mean Shift
CVS	9807	9807	5744	715	2622	9788	9799	9802
Lowe's	1741	1741	972	99	499	1741	1741	1741
Walmart	5746	5743	2793	341	1640	5723	5725	5735

Table 5. Hypothesis test results to determine if differences in drive times are statistically significant. Retain indicates that we failed to reject the NULL hypothesis, indicating that there is no difference in drive time. Reject indicates that we reject the NULL hypothesis, indicating that there is a difference in drive time. A * indicates that results were computed with less than 200 locations due to a cluster centroid being further than 5 miles away.

	Census vs. Traditionally clustered			Census vs. Recursively reclustered		
	AP	k-Means	Mean shift	AP	k-Means	Mean shift
CVS	Retain*	Retain*	Retain*	Retain	Retain	Reject
Lowe's	Retain*	Retain*	Retain*	Retain	Reject	Reject
Walmart	Retain*	Retain*	Retain*	Retain	Retain	Reject

95 Walmart locations have no clustered block groups within 5 miles when using affinity propagation. When using k-means 190 CVS, 192 Lowe's, and 187 Walmart locations have no clustered block groups within 5 miles. Finally, for mean shift 157 CVS, 137 Lowe's and 153 Walmart locations do not have a clustered block group within 5 miles. From Table 6, we observe that the 95% confidence interval of the differences are between 20 s to 327 s. Traditional clustering is unsuitable to generate optimal clusters for improving drive time computations, as over 40% of the locations in our sampled dataset have a cluster centroid that is more than 5 miles away.

6.2   Comparison to Recursive Reclustering

We use a paired t-test to determine if the differences in drive times for census block group and recursively reclustered data are significantly different by testing the following hypotheses:

NULL: the mean drive time for census block group data is no different from the mean drive time for recursively reclustered data.

Alternate: the mean drive time for census block group data is different from the mean drive time for recursively reclustered data.

From Table 4, we observe that all CVS, Lowe's, and Walmart locations contain at least one census block group within 5 miles. Our recursively reclustering approach shows 0.4% data loss, with at most 19 out of 9,807 total CVS location and 23 out of 5,746 total Walmart locations having no recursively reclustered block group within 5 miles.

From Table 5, we see no statistically significant differences in drive time for all three datasets using affinity propagation. We see no statistically significant difference in drive times for Walmart and CVS for k-means, while a significant difference in drive time was observed for Lowe's, the difference in drive times is under 30 s. While we see a statistically significant difference in drive time for mean shift, the difference is less than 30 s and thus practically insignificant.

Table 6. 95% confidence interval of difference in drive times. The values for the confidence interval are provided in seconds. A ∗ indicates that results were computed with less than 200 locations due to a cluster centroid being further than 5 miles away.

	Census vs. Traditionally clustered			Census vs. Recursively reclustered		
	AP	k-Means	Mean shift	AP	k-Means	Mean shift
CVS	$[-39.9, 45.5]$∗	$[-104.1, 108.9]$∗	$[-38.0, 24.4]$∗	$[-8.0, 12.3]$	$[-15.8, 13.0]$	$[10.1, 20.2]$
Lowe's	$[-45.5, 34.8]$∗	$[-327.2, 239.6]$∗	$[-22.2, 19.2]$∗	$[-12.6, 6.0]$	$[8.6, 28.3]$	$[11.6, 20.3]$
Walmart	$[-61.5, 21.9]$∗	$[-271.0, 80.3]$∗	$[-137.3, 41.4]$∗	$[-5.3, 13.6]$	$[-0.3, 26.9]$	$[4.0, 20.2]$

From Table 6, we observe that our recursive reclustering approach yields drive times within 30 s of the actual drive time obtained from the census block group data. For the 200 random CVS Pharmacy locations, we computed drive times to 34,888 locations when using the census data, but only 5,516 when using affinity propagation, 4,990 using k-means, and 32,593 using mean shift. For the 200 random Lowe's Home Improvement locations we compute 21,249 drive times using the census data, but only 3,604 using affinity propagation, 3,202 using k-means, and 19,105 using mean shift. Finally, for the 200 random Walmart locations we computed drive times to 17,876 locations using the census data, and only 2,997 locations using affinity propagation, 2,710 locations using k-means, and 16,528 locations using mean shift.

Our recursively reclustered affinity propagation and k-means approach provides a 6× reduction in the number of computations with no practically perceivable difference in the drive times. For example, as shown in Fig. 8 for a random location denoted by the diamond symbol and located at coordinates $(41.766458, -72.677643)$, we generate 253 potential customers groups in a 5 mile bounding box using the census block group data with an average drive time of 10 min 14 s. Our approach generates 33 clustered customer groups with an average drive time of 10 min 5 s [18].

6.3   Differences in Actual and Recursively Reclustered Drive Times

Drive time differences between the census and recursively reclustered data arise due to the number of points within the trade area for a given location. In Fig. 9, we observe that for all three clustering algorithms, the differences between actual drive times and recursively reclustered drive times increase as the number of block group clusters decreases. In sparsely populated areas, census block groups are separated by larger spatial distances. In such areas, we observe drive time differences up to 2 min. In densely populated areas, where census block groups are spatially closer we observe lower drive time differences, many of which are within 30 s. Residents in sparsely populated areas are more likely to be accepting of slightly longer differences in drive times due to the inherent sparsity of resources.

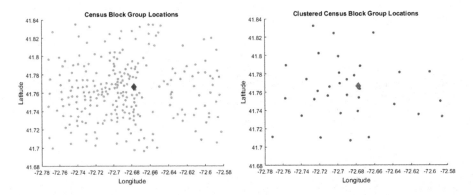

Fig. 8. Effect of clustering on reducing the number of drive time computations in a urban location, such as Hartford, CT. The diamond indicates a proposed location, and the circles indicate block groups. The figure on the left shows the raw census block group data, while the figure on the right shows the clustered block group data [18].

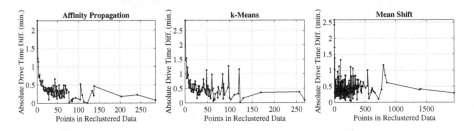

Fig. 9. Drive time differences measured in minutes for census vs. clustered census data using affinity propagation, k-means, and mean shift. Drive time differences reduce as the number of clustered points in the neighborhood of a proposed location increases.

7 Discussion

In this work, we perform a comparison of our approach on constrained recursive reclustering introduced in our prior work [18] for affinity propagation, on two additional clustering algorithms, i.e., k-means and mean shift, and on a variety of site location datasets. Unlike the affinity propagation algorithm used in our original work, both k-means and mean shift require the pre-specification of user-defined parameters, such as cluster count for k-means and kernel bandwidth for mean shift. We use the approach of silhouette score computation to select an optimal cluster count in k-means, while we use the 30th percentile of pairwise distances as the kernel bandwidth in mean shift. Our recursive reclustering approach provides reductions of 81.2%, 83.4%, and 10.2% for affinity propagation, k-means, and mean shift respectively when compared to the 220,334 census block groups. Using 200 randomly sampled locations each from Lowe's, CVS, and Walmart, we show that compared to the original block groups there is no statistically significant difference in drive time computations when using clusters generated by constrained recursive reclustering with affinity propagation for any of the three businesses, and with k-means for CVS and Walmart. While statistically significant differences are obtained with k-means for Lowe's and with mean shift for all three businesses, the differences are negligible, with the mean difference within each location set being within 30 s.

While we do not use the haversine distance to compute the drive time itself, we do use it as a heuristic in the constraint set in Subsect. 3.3 to perform cluster splitting. In future, we will perform statistical comparison on using the haversine distance heuristic and on using higher resolution geographical data such as locations of natural relief, traffic patterns, and differences in speed limits of local roads to perform cluster splitting. In this work, we use pre-defined thresholds of 10 for the cluster point counts, 20,000 individuals for the population cutoff, and 5 miles for the distance bound. In future, we will perform statistical analyses with a range of population and distance thresholds. Currently, our approach performs splitting of clusters till they satisfy the constraint set. However, in the event that small clusters are created during the initial clustering, as in the mean shift approach, it is possible for the small clusters to be re-grouped into a larger cluster that still satisfies the constraint set. In future work, we will perform bottom-up cluster regrouping to obtain the minimal number of clusters that just meet the constraint set. Future work will also include a comparison of computation performance and statistical analysis on drive time computation results for constrained recursive reclustering on other groups of clustering algorithms such as spectral and hierarchical clustering.

References

1. Aras, H., Erdoğmuş, Ş., Koç, E.: Multi-criteria selection for a wind observation station location using analytic hierarchy process. Renewable Energy **29**(8), 1383–1392 (2004)

2. Athas, W.F., Adams-Cameron, M., Hunt, W.C., Amir-Fazli, A., Key, C.R.: Travel distance to radiation therapy and receipt of radiotherapy following breast-conserving surgery. JNCI **92**(3), 269–271 (2000)

3. Banaei-Kashani, F., Ghaemi, P., Wilson, J.P.: Maximal reverse skyline query. In: Proceedings of ACM SIGSPATIAL, pp. 421–424 (2014)

4. Blanchard, T., Lyson, T.: Access to low cost groceries in nonmetropolitan counties: large retailers and the creation of food deserts. In: Measuring Rural Diversity Conference Proceedings, pp. 21–22, November 2002

5. Bradley, P., Bennett, K., Demiriz, A.: Constrained k-means clustering. Microsoft Research, Redmond, pp. 1–8 (2000)

6. Branas, C.C., et al.: Access to trauma centers in the United States. JAMA **293**(21), 2626–2633 (2005)

7. Carr, B.G., Branas, C.C., Metlay, J.P., Sullivan, A.F., Camargo, C.A.: Access to emergency care in the United States. Ann. Emerg. Med. **54**(2), 261–269 (2009)

8. Çebi, F., Otay, I.: Multi-criteria and multi-stage facility location selection under interval type-2 fuzzy environment: a case study for a cement factory. IJCIS **8**(2), 330–344 (2015)

9. US Census: 2010 us census block group data (2010). http://www2.census.gov/geo/docs/reference/cenpop2010/blkgrp/CenPop2010_Mean_BG.txt

10. Chen, L., et al.: Bike sharing station placement leveraging heterogeneous urban open data. In: Proceedings of ACM Ubicomp, pp. 571–575 (2015)

11. Comaniciu, D., Meer, P.: Mean shift: a robust approach toward feature space analysis. IEEE Trans. Pattern Anal. Mach. Intell. **24**(5), 603–619 (2002)

12. Farber, S., Morang, M.Z., Widener, M.J.: Temporal variability in transit-based accessibility to supermarkets. Appl. Geogr. **53**, 149–159 (2014)

13. Frey, B.J., Dueck, D.: Clustering by passing messages between data points. Science **315**(5814), 972–976 (2007)

14. Ghaemi, P., Shahabi, K., Wilson, J.P., Banaei-Kashani, F.: Optimal network location queries. In: Proceedings of ACM SIGSPATIAL, pp. 478–481 (2010)

15. Ghaemi, P., Shahabi, K., Wilson, J.P., Banaei-Kashani, F.: Continuous maximal reverse nearest neighbor query on spatial networks. In: Proceedings of ACM SIGSPATIAL, pp. 61–70 (2012)

16. Goodman, D.C., Fisher, E., Stukel, T.A., Chang, C.h.: The distance to community medical care and the likelihood of hospitalization: is closer always better? Am. J. Public Health **87**(7), 1144–1150 (1997)

17. Google: Google Maps Distance Matrix API (2017). https://developers.google.com/maps/documentation/distance-matrix/

18. Gwinn, D., Helmick, J., Banerjee, N.K., Banerjee, S.: Optimal estimation of census block group clusters to improve the computational efficiency of drive time calculations. In: GISTAM, pp. 96–106 (2018)

19. Jiao, J., Moudon, A.V., Ulmer, J., Hurvitz, P.M., Drewnowski, A.: How to identify food deserts: measuring physical and economic access to supermarkets in King County, Washington. Am. J. Public Health **102**(10), e32–e39 (2012)

20. Kahraman, C., Ruan, D., Doğan, I.: Fuzzy group decision-making for facility location selection. Inf. Sci. **157**, 135–153 (2003)

21. Karamshuk, D., Noulas, A., Scellato, S., Nicosia, V., Mascolo, C.: Geo-spotting: mining online location-based services for optimal retail store placement. In: Proceedings of ACM SIGKDD, pp. 793–801 (2013)

22. Kuo, R., Chi, S., Kao, S.: A decision support system for locating convenience store through fuzzy AHP. Comput. Ind. Eng. **37**(1), 323–326 (1999)

23. Li, Y., Zheng, Y., Ji, S., Wang, W., Gong, Z., et al.: Location selection for ambulance stations: a data-driven approach. In: Proceedings of ACM SIGSPATIAL, p. 85 (2015)
24. Lloyd, S.: Least squares quantization in PCM. IEEE Trans. Inf. Theory **28**(2), 129–137 (1982)
25. Love, R.F., Morris, J.G.: Mathematical models of road travel distances. Manage. Sci. **25**(2), 130–139 (1979)
26. Nallamothu, B.K., Bates, E.R., Wang, Y., Bradley, E.H., Krumholz, H.M.: Driving times and distances to hospitals with percutaneous coronary intervention in the United States. Circulation **113**(9), 1189–1195 (2006)
27. Nattinger, A.B., Kneusel, R.T., Hoffmann, R.G., Gilligan, M.A.: Relationship of distance from a radiotherapy facility and initial breast cancer treatment. JNCI **93**(17), 1344–1346 (2001)
28. Park, H.S., Jun, C.H.: A simple and fast algorithm for k-medoids clustering. Expert Syst. Appl. **36**(2), 3336–3341 (2009)
29. Pedregosa, F., et al.: Scikit-learn: machine learning in Python. J. Mach. Learn. Res **12**, 2825–2830 (2011)
30. Qu, Y., Zhang, J.: Trade area analysis using user generated mobile location data. In: Proceedings of International Conference on World Wide Web, pp. 1053–1064. ACM (2013)
31. Rokach, L., Maimon, O.: Clustering methods. In: Maimon, O., Rokach, L. (eds.) Data Mining and Knowledge Discovery Handbook, pp. 321–352. Springer, Boston (2005). https://doi.org/10.1007/0-387-25465-X_15
32. Rousseeuw, P.J.: Silhouettes: a graphical aid to the interpretation and validation of cluster analysis. J. Comput. Appl. Math. **20**, 53–65 (1987)
33. Statista: Total number of Walmart stores worldwide from 2008 to 2018 (2018). https://www.statista.com/statistics/256172/total-number-of-walmart-stores-worldwide/
34. Tzeng, G.H., Chen, Y.W.: The optimal location of airport fire stations: a fuzzy multi-objective programming and revised genetic algorithm approach. Transp. Plan. Technol. **23**(1), 37–55 (1999)
35. Tzeng, G.H., Teng, M.H., Chen, J.J., Opricovic, S.: Multicriteria selection for a restaurant location in Taipei. Int. J. Hosp. Manage. **21**(2), 171–187 (2002)
36. Wagstaff, K., Cardie, C., Rogers, S., Schrödl, S., et al.: Constrained k-means clustering with background knowledge. In: ICML, vol. 1, pp. 577–584 (2001)
37. Wang, F., Chen, L., Pan, W.: Where to place your next restaurant?: Optimal restaurant placement via leveraging user-generated reviews. In: Proceedings of ACM CIKM, pp. 2371–2376 (2016)
38. Wang, Y., Jiang, W., Liu, S., Ye, X., Wang, T.: Evaluating trade areas using social media data with a calibrated huff model. ISPRS Int. J. Geo-Inf. **5**(7), 112 (2016)
39. Xiao, X., Yao, B., Li, F.: Optimal location queries in road network databases. In: IEEE ICDE, pp. 804–815 (2011)
40. Xu, M., Wang, T., Wu, Z., Zhou, J., Li, J., Wu, H.: Demand driven store site selection via multiple spatial-temporal data. In: Proceedings of ACM SIGSPATIAL, p. 40 (2016)
41. Yang, J., Lee, H.: An AHP decision model for facility location selection. Facilities **15**(9/10), 241–254 (1997)
42. Yong, D.: Plant location selection based on fuzzy topsis. Int. J. Adv. Manuf. Technol. **28**(7), 839–844 (2006)

43. Yu, Z., Tian, M., Wang, Z., Guo, B., Mei, T.: Shop-type recommendation leveraging the data from social media and location-based services. ACM TKDD **11**(1), 1 (2016)
44. Yu, Z., Zhang, D., Yang, D.: Where is the largest market: ranking areas by popularity from location based social networks. In: IEEE UIC/ATC, pp. 157–162 (2013)

Emergence of Geovigilantes and Geographic Information Ethics in the Web 2.0 Era

Koshiro Suzuki[(⊠)]

Faculty of Humanities, University of Toyama, 3190 Gofuku,
Toyama 930-8555, Japan
lichthoffen@hotmail.com

Abstract. The current technical evolution has enabled GIS as a tool for social participation, empowerment, and public involvement. Public citizens become acclimatized to voluntarily participate in regional policy planning, local governance and crisis mapping. However, it also brings the consequences that people can casually participate in mapping behaviour without being aware of their position of power in creating geographic information without knowledge of cartography or ethics. Thus, the premise that such net-rooted and undisciplined people do what experts expect of them no longer applies. In this paper, based on the literature on online activism and digital vigilantism, the author introduces the notion of geovigilantism and highlights the necessity of developing geographic information ethics for PGIS in the Web 2.0 era by referring to two types of recent online cyberbullying incidents. Because technology-aided ubiquitous mapping is difficult to see or grasp, especially for those not educated and trained to see it, these advances prompt people to lower technical and ethical barriers. Further studies are essential to establish geographic information ethics and address this newly emerging problem.

Keywords: Ubiquitous mapping · Absconditus · PGIS ·
Geographic information ethics · Cyberbullying · Geovigilantism

1 Introduction

Since the 1990s, the possibilities of geospatial analysis in conjunction with GIS have dramatically increased in the context of the consolidation of geostatistical data, high precision GPS, improvement of PC processing capability, and expedited LAN access. Geographers gradually became aware of the magnitude of the social impacts of the GISystem, which became capable of analysing and outputting even personal-level data [1].

Advancements in geospatial information technology (GIT) have increased the necessity of dealing with GIS from an interdisciplinary science perspective to examine the social influence of the innovation and functionality of the system. Although technological advances in GIS triggered a series of debates in the 1990s on its scientific ambiguity and social implications, these 'science wars' led to the establishment of a new discipline, namely GIScience [2, 3].

© Springer Nature Switzerland AG 2019
L. Ragia et al. (Eds.): GISTAM 2018, CCIS 1061, pp. 55–72, 2019.
https://doi.org/10.1007/978-3-030-29948-4_3

One benefit emerging from the debate was the use of GIS as a tool for social participation, empowerment, and public involvement through Public Participation GIS (PPGIS) or Participatory GIS (PGIS). PGIS/PPGIS were largely enabled by GIT-aided ubiquitous mapping and cartography [4–6]. The historical background of PGIS/PPGIS is reviewed in the next section. In contrast, excepting studies on privacy, little attention has been paid to the implications of technology-aided incidents. In this paper, the necessity of developing new geographic information ethics is highlighted by introducing the concept of 'geovigilantism'. In doing so, existing debates in relevant fields are reviewed to clarify what has been overlooked and what should be considered.

2   Rise of PGIS and Internet Privacy

2.1   Technical Progress in GIT

Historically, essential GIT technologies relied on advancements in remote supervision for military purposes. The first earth imagery taken from outer space (thermosphere: 100 km above) was made possible using the technology of the former V-2 rocket engineers after World War 2 during army-led experiments as part of 'Operation paperclip' in 1946 [7]. This served as a launching pad for satellite imagery technology today. Likewise, aerial photography was used for the high altitude carpet bombing of Japan [8] during the Pacific War, although it was initially developed in the US for remote agricultural land monitoring in 1936 [9]. Critical technology related to remote sensing stems from the infrared camera, which was invented by Kodak in 1943. This invention was similarly ground-breaking for military purposes, because even a camouflaged building could be easily penetrated using infrared imaging [10]. In the Gulf War, GPS technologies were employed in combination with precision-guided systems for the surgical bombing of Iraq [11]. These historical facts demonstrate that the essential elements of GIS/GPS were derived from and built for the purpose of telemonitoring. As such, GIS/GPS is predisposed as a tool for remote supervision including hotspot analysis in predictive policing [12, 13] and residency restrictions, which based on Megan's Law in the US [14, 15], restricts ex-sexual offenders from establishing or occupying a residence within 1,000 feet of school-related properties [16]. These spatial restrictions are enabled through the establishment of GPS-based real time telemonitoring and geovisualisation.

Therefore, at the initial stage of ubiquitous mapping, satellite-based geolocation tracking played an important role as a technological platform [17, 18]. In 1972, the maximum resolution of satellite imagery was only 79 m^2 [19], but this is now improving. Specifically, Japan is constructing a self-centred quasi-zenith satellite system (QZSS) [20], for which accuracy has been improved to the mm level of precision.

Accordingly, in the US, the Federal Communications Commission (FCC) mandated in 1996 that all wireless service providers deliver accurate location information of emergency 911 (E-911) callers to public safety answering points (PSAPs), because a large proportion of 911 calls originated from mobile phones [21]. In 2007, the Japanese government also intended to establish a Japanese E-911 system based on the partial revision of the Regulations for Telecommunications Facilities for Telecommunications Business under the jurisdiction of the Ministry of Internal Affairs and Communications. It mandated that all subsequent 3G and newer cellular phone generations carry GPS to make their current locations automatically detectable by the police, fire department, and coast guard emergency call stations [22]. With the technological advancement of the Indoor Positioning System (IPS) [23] and Indoor Messaging System (IMES) [24], human location tracking is becoming geographically seamless and entering our homes.

In the mid-2000s, this situation dramatically changed. Thanks to the consolidation of wireless LAN networks such as Wi-Fi, which was initially established in 2000, hardware became capable of providing portable high-speed Internet access. Moreover, interactive web services such as Mixi, Facebook, Twitter, posting bulletin boards, and Flickr were launched successively throughout the decade. As a result, more casual and unbounded communication between people was enabled via the Internet, and Web 2.0 became a reality. Importantly, when map integration technology (GeoAPI) was put into practical use on the web, it played a fundamental role in this technological progress. Google began to provide Street View and Google Earth in 2006–2007 in addition to Google Maps, which had previously launched in 2005. For Street View, Google started creating a sequential and panoramic cityscape archive, collecting photos through camera-mounted Street View cars and trikes. For Google Earth, Google began exploiting high-resolution satellite imagery taken by Worldview-2 and Worldview-3 commercial satellites operated by DigitalGlobe in 2009 and 2014 respectively [25, 26]. Both services are free of charge for personal browsing. Using these platforms and devices, everyone began geo-tagging and sharing photos and texts on web maps. In summary, it is important to acknowledge that advancements in ubiquitous mapping relied on corroborative military-based tele-monitoring technologies. We now outline the history of PGIS/PPGIS.

2.2   Positive Aspects of PGIS/PPGIS

The technical evolution has enabled public citizens, who were simply receivers of geographic information, to become senders, sharers, and communicators of geographic information using social networking services (SNS) and online mapping devices [27, 28]. Sometimes, these grassroots mappers voluntarily participate in regional policy planning and local governance, which are referred to as volunteered geographic information (VGI) [29], neogeography [27], and bottom-up GIS [30]. One positive aspect of these technological innovations is the rise of PGIS [31], a recent synonym of PPGIS. Essentially, PPGIS conveyed nuances of community development and landscape ecology and the aspiration to empower those disadvantaged and marginalised through GIS technology [32, 33]. Likewise, despite the breadth of relevant areas, most PPGIS projects 'have evoked interest from researchers and practitioners' (Sieber, 2006: 492) [33]. This tendency can perhaps be traced to the fact that initially, GIS was intended as a regional spatial decision support tool. In fact, in a ground-breaking study, Peluso (1995) [34] used collaborative sketch mapping to compare indigenous and contesting occupancy rights and forest territories customarily claimed or managed by local people with official forest mapping by government forestry planners. This bottom-up and collaborative mapping, termed counter-mapping, established GIS as a tool for empowerment in geospatial planning and decision making [35, 36].

The target people were not only disadvantaged and indigenous, but did not have exit rights because of the effects of heavily deployed GIT [37]. This disadvantage provided ethical justification to researchers and planners, who became involved in empowerment activities. Actually, numerous case studies focus on marginalised people, regions, and countries [38–40].

Geospatial technologies and PPGIS are new to the people in such areas, at least in the early phase, meaning that an anisotropic (interactive but not bidirectional) social relationship develops in the project between those who empower/manage and those who are empowered/participate in terms of jurisdictional authority. Sometimes, these elite-driven mapping projects encompass social and environmental change in communities and cause unintended consequences such as increased conflicts among villagers, loss of the indigenous conception of space, and increased privatisation of land [41]. Therefore, ethics related to PGIS were developed to provide a code of professional ethics. For example, based on scholarly debates in the early and mid-1990s, Rambaldi et al. [42] formulated a guideline for 'the obligation of the individuals to make their best judgment' (Ibid., 108) regarding good practice and PGIS ethics. The guide principles stated that users should (1) be open and honest, (2) obtain informed consent, and (3) do their best to recognise that they are working with socially differentiated communities and their presence will not be politically neutral. As the clauses indicate, the assumed users of the guidebook were people intending to empower these communities.

With fair-use and the communisation of geographic information, some open source GIS programs and web-based open-source mapping platforms were established [43, 44]. This progress enabled grassroots mappers to use GIT in the post-disaster construction and damage repair process by digitising satellite imagery of the afflicted areas on OpenStreetMap to find ways around damaged roads [45]. These crisis-mapping actions demonstrate the efficacy thereof in the aftermath of the 2011 earthquake off the Pacific coast of Tōhoku [46]. Generally, cartographers positively interpret the actions as a people-powered, net-rooted, undisciplined, alternative, and Dionysian way of mapping [27, 47]. Sometimes, such empowered mappers challenge the hidden 'places and facilities, including a panoply of military installations, sites relating to state security, policing and prisons, and increasingly "strategic" national assets and infrastructures' (Perkins and Dodge, 2009: 546) [48] to be revealed and shared through GIT.

Recently, several researchers have begun focusing on capturing unintended disseminated geospatial components (e.g. geotagged pictures and text describing a location) in the tweets and entries of SNS users as potential sources of geographic analyses [49]. Stefanidis et al. termed such digital footprints ambient geospatial information (AGI), in contrast to VGI [50]. Some empirical studies confirm that involuntarily released big data can be used for geospatial analyses [51, 52], thereby expanding the possibilities of GIS.

2.3 Negative Aspects of PGIS

On the other hand, potential threats also stem from advancements. Numerous studies on GIScience focus on the issue of Internet privacy. As noted, GIS/GPS derives from and is built for the purpose of telemonitoring, meaning they were developed as tools for remote supervision. Therefore, critics who emphasised the potential threats of the technical progress of GIS tend to focus on privacy.

Boyd and Ellison's [53] review of studies dealt with privacy issues inherent in the online environment, summarising these as (1) damaged reputation due to rumours and gossip, (2) unwanted contact and harassment or stalking, (3) surveillance-like structures because of backtracking functions, (4) use of personal data by third parties, and (5) hacking and identity theft [54]. Previous studies on SNS privacy issues focused on ethical questions involving the remote monitoring of users by the service provider. In addition, the invasion of privacy and surveillance of geographic space as an exercise of public power are subjects of considerable discussion in GIScience [19]. Although the discussion on mapping is still limited, scholars have coined the term 'geosurveillance' [55] to spark a critical discussion on the potential risks of privacy infringement through the aggregation of users' attributes and location information collected by public authority and SNS providers. Although the development of Information and Communications Technology (ICT) enables people to share geographic information in a friendlier manner, through geosurveillance, users remain under tight scrutiny [9]. Dobson and Fisher [56] defined the term 'geoslavery' as 'a practice in which one entity, the master, coercively or surreptitiously monitors and exerts control over the physical location of another individual, the slave' (Ibid., 48). Many scholars metaphorically refer to the 'big brother' motif in George Orwell's famous novel '1984' to describe the power and position of a master [57, 58], and Bentham's panopticon for the systems and techniques of monitoring [59, 60]. Although many studies are extremely conscious of the potential risks of geosurveillance by public powers, their discussions on privacy infringement at the personal level lack diversity.

There are many empirical studies on individual victims via SNS. For instance, Gross and Acquisti [61], a classic empirical study on SNS profiles, found that 89% of users use their real names in their Facebook profiles, and 61% use identifiable information in their posts. Furthermore, Jones and Soltren [62] confirmed that 62% of student users did not configure privacy settings, despite that 74% of them knew about Facebook privacy options. They also pointed out that 70% of Facebook users posted personal information. In other words, they cannot effectively protect their privacy, despite caring about the leakage thereof, in what Barnes [63] refers to as the 'privacy paradox'. Other studies identified another rationale, namely that the tendency to inadequately protect one's private information was the consequence of exhibitionistic motives [64, 65]. These studies demonstrate the vulnerability of potential victims, who are not protected against anonymous third party offenders on the Internet. However, not much is revealed about the offenders. Previous studies on the negative effects of ubiquitous mapping emphasise the risks of privacy infringement through public power or criminological studies through SNS. However, in the Web 2.0 era, the panoptic one-to-many relationship becomes the 'many-surveying-the-many' situation, described by Shilton [66] as little brothers and by Rose-Redwood [67] as the omnopticon. In these views, the progress of PGIS may encompass the participatory panopticon and total loss of privacy [68]. Kawaguchi and Kawaguchi [69] rephrased the omnopticon as 'paradoxical others' to describe the feeling of discomfort upon being disclosed on Google Street View. Liberally interpreted, these views suggest that an omnoptic mutual surveillance environment restrains and intermediates people from deviant behaviours as an unseen hand of God.

While studies rarely concentrate on online offenders in the context of GIT and PGIS, those on cyberbullying or cyberstalking focus on adolescent students and offer numerous suggestions regarding online offenders [70, 71]. These studies can be categorised into two types. The first group deals with sex crimes toward adolescents via the Internet in a criminological context. Alexy et al. [72] asked 756 university students about their cyberstalking experiences, finding that males were more likely to have been cyberstalked, most likely by a former intimate partner. Likewise, Sheridan and Grant [73] conducted an online questionnaire survey of 1,051 self-defined stalking victims about their experience of victimisation. They concluded that cyberstalking did not fundamentally differ from traditional, proximal stalking; that online harassment did not necessarily hold broad appeal to stalkers; and that most stalkers targeted ex-intimates.

In research targeting younger school-aged children, pupils and students were considered victims. The Crime against Children Research Center at the University of New Hampshire conducted a series of youth Internet safety surveys every five years since 2005, including the National Juvenile Online Victimization Study, the only research investigating Internet-initiated sex crimes. According to the survey, 73% of victims who had face-to-face sexual encounters with offenders did so more than once, because most loved or were affectionate toward their offenders [71]. Despite the victim's agreement, their acts arouse research concern regarding the linkage of sex crimes to juveniles in the legal context. In summary, apart from the newness of the method used, offenders must establish spatial proximity to some extent when seeking physical contact. Therefore, by nature, cyberbullying and cyberstalking are geographically constrained.

The second group of studies on cyberbullying focus on in-school online bullying. These studies are characterised as behavioural and educational, rather than criminological. Kowalski and Limber [74] conducted questionnaire surveys of 3,767 elementary and middle school students in the US, finding that 11% had been bullied online in the last couple of months, and 4% had electronically bullied someone else. They further reported that the most common means of bullying were through instant messaging, chat messaging, and e-mail. Furthermore, the form of bullying differed according to gender. Many studies reported that males tend to engage more in direct and physical forms of aggression, while females use insults or ostracise others, similar to offline or face-to-face in-school bullying [75–77]. In addition, males are more likely to be both offenders and victims [78–80]. Although they investigated the usage of online devices and tools used when bullying, this act is based on students' in-school social relationships and geographic proximity. Therefore, excepting the novelty of the medium, cyberbullying in this context is neither an Internet specific nor new phenomenon. As mentioned, current studies that focus on GIT-aided ubiquitous mapping generally portray users as naïve, well-intentioned, and cooperative people. However, these people can also employ and utilise the new tools for their purposes when they become available. In fact, previous studies demonstrated that terrorists and criminals use GIT to implement their plans [81–83]. As demonstrated by studies on cyberbullying or cyberstalking, new technology can be used in both good and bad ways. Before the Web 2.0 era, most people who could create and manage maps were knowledgeable experts who had been educated and had internalised codes of professional ethics. In contrast, in ubiquitous mapping, people can casually participate in mapping behaviour without being aware of their position of power in creating geographic information without knowledge of cartography or ethics.

Thus, the premise that net-rooted, undisciplined, alternative, Dionysian people do what experts expect of them no longer applies.

In the next section, two online mobbing cases are presented to examine the possibilities not yet discussed in preceding contributions. With the benefit of expertise in the related area of digital activism, the notion of geovigilantism is introduced and the necessity of developing geographic information ethics in PGIS studies highlighted in the concluding section.

3 Digital Vigilantism and Geovigilantes

3.1 Online Activism and Digital Vigilantism

As noted, other than research on cyberbullying or cyberstalking, studies on online offenders is lacking in the field of GIScience. The current study addresses this gap by examining related areas, namely online activism and vigilantism.

Online activism and vigilantism focus on online collective action. Counter-mapping played a constructive role in initial progress in PPGIS as an empowerment tool for people who were politically marginalised. Similarly, online activism is a politically motivated public movement against the power elite [84]. Therefore, an early ideological form of online activism is 'Hacktivism'. Hacktivism is defined as a 'form of civil disobedience, which unites the talents of the computer hacker with the social consciousness of the political activist' (Manion and Goodrum, 2000: 14) [85]. Because the practice relies on the skills of hackers, it embraced the ethical contradiction of the 'ends-justify-the-means'.

Considering the motives of the agents involved in digital vigilantism, Suler [86] proposed the 'online disinhibition effect', which comprises six factors: dissociative anonymity, invisibility, asynchronicity, solipsistic introjections, dissociative imagination, and the minimisation of authority. Online, people 'do not have to worry about how others look or sound in response to what they say' and 'in the case of expressed hostilities or other deviant behaviors, the person can avert responsibility for those behaviors, almost as if superego restrictions and moral cognitive processes have been temporarily suspended from the online psyche' (Ibid., 322). The author classifies disinhibition in terms of its benign (positive) and toxic (negative) effects. Regarding the latter, numerous follow-up experimental studies have been conducted to identify contributing factors. For example, Lapidot-Lefler and Barak [87] studied the impact of three typical online communication factors on inducing the online disinhibition effect: anonymity, invisibility, and lack of eye contact. They determined that the lack of eye contact was the chief contributor to the negative effects of online disinhibition, contrary to Suler's expectations that anonymity would be the principle factor.

Furthermore, this effect makes people become complicit in citizen-led Internet/cyber/online/digital violence and surveillance, referred to as 'vigilantism' [88, 89]. Notable case studies of online vigilantism deal with recent Chinese cyberspace incidents. Since the government opened cyberspace to citizens in 1997, thousands of netizens began engaging in cyberspace governance. Although their acts of vigilantism 'often explicitly support the central government while attacking local officials, institutions, etc. or even other netizens perceived to have broken the unwritten rules of Chinese cyberspace' (Herold, 2008: 26) [90], their skills regarding searching for personal information—known as a 'human flesh search' [91]—confirmed the potential power and threat to privacy of 'doxing' and demonstrated many-to-many mutual surveillance online.

Although the scope of these studies did not include the impact of progress in GIT on PGIS, collective vigilantes arose from the emergence of the Web 2.0 environment. Since the environment also technologically underpins the existence of PGIS, findings on online vigilantism can be assimilated into a new form of geographic information ethics. When ubiquitous mapping aided by GIT and Web 2.0 plays an essential role in digital vigilantism, it is defined as geovigilantism. Likewise, people who actively participate therein are considered geovigilantes. In the next section, two recent online incidents that can be understood as an expression of geovigilantism are reviewed.

3.2   Individual Online Peepers

On 20 February 2015, a murder occurred in Kanagawa prefecture, Japan. The 13-year-old victim had tried to withdraw from the perpetrators' circle, and was found bound and stabbed to death by three juvenile criminals. The case received much media coverage, because of the atrocity of the crime and ages of the accused.

However, the case is memorable not only because of the savagery, but also in the context of the present paper. The then 15-year old podcaster, whose handle name was Noeru (Noël), located and found the chief culprit's family home, and webcast it globally. Figure 1 is a screenshot of the delivered movie (now deleted) showing the symbolic composition of a journalist holding out a microphone to a nameless boy as seen from behind him. The figure demonstrates that even a boy can compete with the professional media in terms of competence in information transmission. Noeru (and other mappers) could determine a location by specifying aggregated place names and utilising Google Street View to find the same exterior appearance of the home broadcast by the mass media to detect the exact target location (termed 'dataveillance') [92]. Why did the boy, who was not involved, do this? It can be inferred that he aspired to fame and increased advertisement revenue, even if he becomes considered an online 'weirdo'.

Fig. 1. Screenshot of the podcast movie (*now deleted) [83, 93].

Such podcasters are everywhere, biding their time to gain overnight fame. On 31 December 2017, an American Youtuber identified as 'Logan Paul' posted a 15-min video showing him and his friends going to 'Aokigahara', a well-known suicide spot at the bottom of Mt. Fuji in Japan, where they fortuitously found a dead body hanging from a tree. Although the comment column was inundated with accusing reactions and he was coerced into withdrawing the post and uploading an apology, the fuss contributed to him becoming a 'buzzfeed', which watched by thousands of people and covered by the media, generated enormous revenue for him (Fig. 2).

Fig. 2. Screenshot of an ABC newscast posted on YouTube reporting Logan Paul's incident [94].

3.3   Private Sanctions by Collective Mobs

On 15 May 2012 in Hachioji, Tokyo, an elementary schoolchild was on his way home from school. He was surrounded by two junior high school students, who were video recording the incident with a cell phone. The two adolescents found a pretext to quarrel with the boy, forcing him to move backward and whimper in fear. The adolescents then uploaded the movie file on their YouTube account for their own pleasure (Yomiuri Online, 21 July 2017).

Immediately after the upload, the URL was disseminated on the Internet through SNS, and appeared on the famous online bulletin board 2channel with a fusillade of accusations. An anonymous person promptly created a portal site using @wiki, a free rental wiki maintained by a limited liability company, Atfreaks (Fig. 3). The website served as a 'traffic cop', directing thousands of seekers to the appropriate information. As the sub-domain name /dqntokutei/ shows, the creator of the domain cared less about right or wrong, but about tokutei (identifying) the dqn (an argot for 'homeboys') who deserved to be sanctioned.

Subsequently, thousands of Internet users (mostly 2channel viewers) voluntarily began Googling information about the captured location and analysing previously uploaded files on the YouTube account. The power of collective intelligence was used to pinpoint the filmed location by scoping distinctive landmarks captured in the setting and comparing them with images on Google Street View. The uniform of the perpetrators also revealed the school they attended. Likewise, amateur investigators examined the contents of previously uploaded movies, finding that the faces of the uploaders and their neighbourhoods were visible in some of the files. These online droves dataveillanced all information published online, identified two nameless targets, and privately sanctioned them through complaints to the school and police. Five years after this initial burst of enthusiasm, the portal site remains on the Internet, exposing the faces and locations of involved individuals to the public gaze.

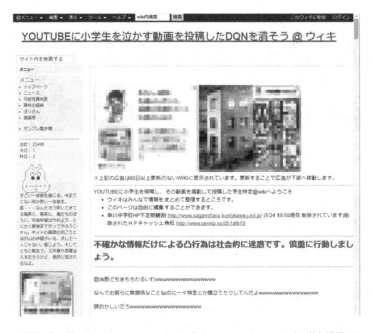

Fig. 3. The top page screenshot of the promptly created wiki [95].

4   Discussion and Conclusion

In 1495 AD in medieval Germany, Ewiger Landfriede was passed by Maximilian I, German king and emperor of the Holy Roman Empire, prohibiting Fehde (the duel) as a self-help right to take vengeance. This was the first time in European history that a Reichskammergericht, the Supreme Court, was established [96]. This event demonstrates that the modern concept of law and justice would not be possible without the consignment of individual rights of vengeance to the public power. Five hundred years and a few decades later, innovation in GIT is prompting the resurgence of this pre-modern principle in a modernised way.

Recently enabled ubiquitous mapping based on the Web 2.0 environment is making it possible for people to create and use geographic information anywhere at any time without advanced map-use skills [4–6]. However, as implied in the meaning of the word, ubiquitous means being omnipresent, like air, health, and water, which are all taken for granted.

In Latin, an antonym of ubiquitous is absconditus, which means being hidden or difficult to see or grasp [97]. Although air is everywhere, its existence is overlooked because of its ubiquitous nature. Likewise, in a ubiquitous mapping situation, its presence becomes difficult to see or grasp, especially for people not educated and trained to 'see' it. As demonstrated by the cases in this study, technological advances enable people to participate by lowering technical, cognitive, and ethical barriers.

For now, PGIS studies examine the sunny side of advancements in GIT-aided ubiquitous mapping. Relevant studies on online ethics rely on spontaneously arising equilibrium innervated by mutual surveillance among the people involved [67, 69]. However, this view of this exponential technological advancement is over-optimistic and ingenuous. GIS is merely a device and tool, and people can use new technologies in both good and bad ways.

Although the potential threat of GIT in GIT-aided ubiquitous mapping and cartography has been highlighted [4–6], little attention has been paid to the implication of technology-aided incidents. In this paper, existing debates in relevant fields were critically summarised to clarify the ethical challenges in PGIS. Advocating the concept of geovigilantism, two types of geovigilantes were identified.

Of the two types identified in the present study, the latter collective mobbing case demonstrates the typical digital vigilantism motive. In the web 2.0 environment, incentive cues become ad hoc, so that no-one can state a definite date and reason for inundating an SNS account with comments. Likewise, in contrast to conventional digital vigilantism, behaviourally, collective mobs are likely curious bystanders, not people disobeying power and authority structures. In fact, while blogs and bulletin boards are inundated (*Enjou*: burnt, in Japanese online slang) with the choleric comments of collective mobs, many ASCII arts (AAs) also joke about the victims, sparking further reactions. In Japan, AAs often attached to the burnt spot, as seen in Fig. 4, which portrays the event as a 'Matsuri' (festival). These AAs demonstrate that mobs do not always gather out of disobedience or a sense of equity. Thus, the recent rapid progress of GIT has dramatically enhanced the ability of geosurveillance among collective mobs. Once someone has been targeted, mobs can detect the who the nameless person is by taking the target's cues from the location. Tracking the target's home may harm his/her social existence in terms of confidence, relationships, and status, regardless of whether he/she is a private citizen. Therefore, establishing geographical tracking technology at the personal level constitutes an essential part of digital vigilantism, but in a new way.

In the Web 2.0 environment, the rapid progress of GIT is further complicating matters. While previous digital vigilantism focused on collective mobbing and motives based on disobedience, through GIT-generated ubiquitous mapping, even a junior high school student can determine the exact location of a nameless person anywhere in the world and broadcast it globally. This power and the consequences thereof equal those of the conventional mass media, despite the absence of a code of ethics. As this case demonstrated, collectiveness is not essential in Web 2.0 vigilantism. In other words, an emerging population is asserting itself for fun, revenue, or exhibitionism, while others behave out of anger or disobedience for their own justice or equity, as described in the extant literature on digital vigilantism. A previous article by the author contended terming these geovigilantes a Cyber-COP or Casual Offenders for Pleasure [83].

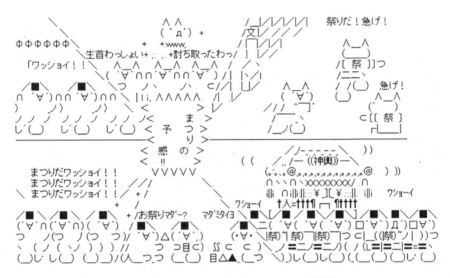

Fig. 4. ASCII art representing a 'Matsuri' [98].

Unlike previously determined online activists and vigilantes, many are juveniles. They cannot legally be held accountable and not be fully aware of the outcomes. Again, important is that even boys and girls can compete with the professional mass media in terms of information gathering and broadcasting abilities. Therefore, further studies should develop geographic information ethics to address this newly emerging problem, and we should be aware of the social role of geography education. Even though geography education cannot deter single-minded fanaticism, it may prevent juveniles from inappropriate behaviour stemming from a sudden impulse or simple ignorance.

Comparative jurisprudence and information ethics are also important. Privacy has been discussed in the field of information ethics since the 1970s. It is now understood as a multi-layered concept, including the right of choice for privacy options [99] and making accountable the infringing party [100]. These aspects have been added to the classic conceptualisation of privacy [101], which regards privacy infringement as stepping into private space. Every legislative system and ethical standard in regions or states place different emphases on these three components.

On 25 May 2018, the Vienna-based NPO noyb.eu filed four complaints of 'forced consent' against Google, Instagram, WhatsApp, and Facebook based on the enforcement of the General Data Protection Regulation (GDPR), because the privacy protections in the GDPR outweigh those in the US [102, 103]. This reflects the differences in legal norms regarding privacy components.

In 1992, an incident occurred in Louisiana. A Japanese student disguised for Halloween was shot dead on the front porch of a house. However, a criminal jury found the defendant innocent [104]. Although the reason for the verdict is not clarified, legally, the student's actions can be regarded as trespassing, which led to his death. The regrettable incident reveals that privacy values in the US are more sensitive to the infringement of private space.

In 2008, Google was involved in a lawsuit 11 months after the launch of Google Street View, because of the psychological damage and loss of asset value caused by a photograph taken from a Street View car. Although the trial dismissed the plaintiff's appeal in February the following year, the plaintiff filed another lawsuit based on unauthorised trespassing on private property. It was decided that Google would accept liability for damages with the payment of $1 [105]. Here, the judgment reflected the American normative consciousness of privacy, in that trespassing on private property was penalised, while dismissing the lawsuit against the infringement of jurisdictional authority [106]. In contrast, in Europe, a similar case was filed against Google in Switzerland. In November 2009, Hanspeter Thür of the Swiss Federal Data Protection and Information Committee (FDPIC) recommended adopting more privacy sensitive countermeasures against inadequate Street View blur processing. This developed into a lawsuit after Google rejected the recommendation [107]. In May 2012, the lawsuit was concluded, and it was decided that foreign companies should apply the rule of Swiss domestic law [108].

Although the Internet is global, each region and state has their own legislative system and ethical standards for artificial and private persons. Therefore, in developing geographic information ethics compatible with the Web 2.0 environment, a deeper appreciation of comparative jurisprudence and information ethics is necessary.

Acknowledgements. This work was supported by JSPS KAKENHI Grant Number JP17H00839. Some parts of this article are based on the following conference presentations conducted by the author: the 63rd Annual Conference of The Japanese Society for Ethics in 2012, the Kyoto Collegium for Bioethics in 2014, the conferences of the Association of Japanese Geographers in 2014 and 2015, and a keynote speech at Hokuriku Geo-Spatial Forum 2017, and GISTAM 2018.

References

1. Miller, H.: Place-based versus people-based GIScience. Geogr. Compass **1**(3), 503–535 (2007)
2. Wright, D.J., Goodchild, M.F., Proctor, J.D.: GIS: tool or science? Demystifying the persistent ambiguity of GIS as "Tool" versus "Science". Ann. Assoc. Am. Geogr. **87**(2), 346–362 (1997)

3. Schuurman, N.: Trouble in the heartland: GIS and its critics in the 1990s. Prog. Hum. Geogr. **24**(4), 569–590 (2000)
4. Morita, T.: A working conceptual framework for ubiquitous mapping. In: Proceedings of XXII International Cartographic Conference, A Courna (2005)
5. Reichenbacher, T.: Adaptation in mobile and ubiquitous cartography. In: William, C., Peterson, M.P., Gartner, G. (eds.) Multimedia Cartography, pp. 383–397. Springer, Heidelberg (2007). https://doi.org/10.1007/978-3-540-36651-5_27
6. Gartner, G., Bennett, D.A., Morita, T.: Towards ubiquitous cartography. Cartogr. Geogr. Inf. Sci. **34**(4), 247–257 (2007)
7. Reichhardt, T.: The first photo from space. Air Space Mag. **24** (2006). https://www.airspacemag.com/space/the-first-photo-from-space-13721411/. Accessed 21 July 2018
8. Office of the Assistant Chief of Air Staff, Intelligence: Records of the U.S. strategic bombing survey entry 48 No. 90.11: Toyama, Japan, security-classified air objective folders 1942-1944, report No. 1-d(18), USSBS Index Section 7. National Archives and Records Administration, Maryland (1944)
9. Monmonier, M.: Aerial photography at the agricultural adjustment administration: acreage controls, conservation benefits, and overhead surveillance in the 1930s. Photogramm. Eng. Remote Sens. **68**(12), 1257–1262 (2002)
10. Monmonier, M.: Spying with Maps: Surveillance Technologies and the Future of Privacy. University of Chicago Press, Chicago (2004)
11. Smith, N.: Real wars, theory wars. Prog. Hum. Geogr. **16**(2), 257–271 (1992)
12. Eck, J.E., Chainey, S., Cameron, J.G., Leitner, M., Wilson, R.E.: Mapping crime: understanding hot spots. National Institute of Justice, Washington D.C. (2005)
13. Anselin, L., Griffiths, E., Tita, G.: Crime mapping and hot spot analysis. In: Wortley, R., Mazerolle, L. (eds.) Environmental Criminology and Crime Analysis, pp. 119–138. Willan, London (2008)
14. Zandbergen, P.A., Hart, T.C.: Reducing housing options for convicted sex offenders: investigating the impact of residency restriction laws using GIS. Justice Res. Policy **8**(2), 1–24 (2006)
15. Levenson, J.S., University, L., Hern, A.L.: Sex offender residence restrictions: unintended consequences and community reentry. Justice Res. Policy **9**(1), 59–73 (2007)
16. Grubesic, T.H., Mack, E., Murray, A.T.: Geographic exclusion: spatial analysis for evaluating the implication of Megan's law. Soc. Sci. Comput. Rev. **25**(2), 143–162 (2007)
17. Zhao, Y.: Mobile phone location determination and its impact on intelligent transportation systems. IEEE Trans. Intell. Transp. Syst. **1**(1), 55–64 (2000)
18. Baba, Y., Kunimitsu, T., Sekine, K., Iwakuni, M., Miyamoto, K.: Emergency report cellular phone, cellular connection switching method and GPS positioning method. U.S. Patent No. 7,127,229. U.S. Patent and Trademark Office, Washington, DC (2006)
19. Armstrong, M.P.: Geographic information technologies and their potentially erosive effects on personal privacy. Stud. Soc. Sci. **27**(1), 19–28 (2002)
20. Harada, Y.: Laying the groundwork for testing routine activity theory at the microlevel using Japanese satellite positioning technology. In: Liu, J., Miyazawa, S. (eds.) Crime and Justice in Contemporary Japan. SSACCJR, pp. 137–151. Springer, Cham (2018). https://doi.org/10.1007/978-3-319-69359-0_8
21. Sayed, A.H., Tarighat, A., Khajehnouri, N.: Network-based wireless location: challenges faced in developing techniques for accurate wireless location information. IEEE Signal Process. Mag. **22**(4), 24–40 (2005)
22. Hino, T.: Developments in location-based services for mobile phones. Mob. Locat. Based Serv. LBS Res. Ser. **2013**, 1–19 (2013)

23. Liu, H., Darabi, H., Banerjee, P., Liu, J.: Survey of wireless indoor positioning techniques and systems. IEEE Trans. Syst. Man Cybern. Part C Appl. Rev. **37**(6), 1067–1080 (2007)
24. Kohtake, N., Morimoto, S., Kogure, S., Manandhar, D.: Indoor and outdoor seamless positioning using indoor messaging system and GPS. In: Proceedings of the International Conference on Indoor Positioning and Indoor Navigation (IPIN 2011), pp. 21–23 (2011)
25. DigitalGlobe: Worldview-2 Datasheet (2016). http://www.digitalglobe.com/sites/default/files/DG_WorldView2_DS_PROD.pdf. Accessed 9 Feb 2016
26. DigitalGlobe: Worldview-3 Datasheet (2016). http://www.digitalglobe.com/sites/default/files/DG_WorldView3_DS_D1_Update2013.pdf. Accessed 9 Feb 2016
27. Turner, A.J.: Introduction to Neogeography. O'Reilly Media, Inc., Sebastopol (2006)
28. Crampton, J.W.: Mapping: A Critical Introduction to Cartography and GIS. Wiley-Blackwell, Malden (2010)
29. Goodchild, M.F.: Citizens as sensors. GeoJournal **69**(4), 211–221 (2007)
30. Talen, E.: Bottom-up GIS. J. Am. Plan. Assoc. **66**, 279–294 (2000)
31. McCall, M.K., Dunn, C.E.: Geo-information tools for participatory spatial planning: fulfilling the criteria for 'good' governance? Geoforum **43**, 81–94 (2012)
32. Chapin, M., Lamb, Z., Threlkeld, B.: Mapping indigenous lands. Annu. Rev. Anthropol. **34**, 619–638 (2005)
33. Sieber, R.: Public participation geographic information systems: a literature review and framework. Ann. Assoc. Am. Geogr. **96**(3), 491–507 (2006)
34. Peruso, N.L.: Whose woods are these? Counter-mapping forest territories in Kalimantan, Indonesia. Antipode **27**(4), 383–406 (1995)
35. Harris, T., Weiner, D.: Empowerment, marginalization and "community-integrated" GIS. Cartogr. GIS **25**(2), 67–76 (1998)
36. Rinner, C.: Argumaps for spatial planning. In: Proceedings of the First International Workshop on TeleGeoProcessing, Lyon, pp. 95–102, 6–7 May 1999
37. Fox, J., Suryanata, K., Hershock, P., Pramono, A.H.: Mapping power: ironic effects of spatial information technology. Particip. Learn. Action **54**(1), 98–105 (2006)
38. Weiner, D., Warner, T., Harris, T.M., Levin, R.M.: Apartheid representations in a digital landscape: GIS, remote sensing, and local knowledge in Kiepersol, South Africa. Cartogr. Geogr. Inf. Syst. **22**, 30–44 (1995)
39. Fox, J.: Siam mapped and mapping in Cambodia: boundaries, sovereignty, and indigenous conceptions of space. Soc. Nat. Resour. **15**(1), 65–78 (2002)
40. Williams, C., Dunn, C.E.: GIS in participatory research: assessing the impact of landmines on communities in North-west Cambodia. Trans. GIS **7**, 393–410 (2003)
41. Fox, J., Suryanata, K., Hershock, P., Pramono, A.H.: Mapping boundaries shifting power: the socio-ethical dimensions of participatory mapping. In: Goodman, M.K., Boykoff, M., Evered, K.T. (eds.) Contentious Geographies: Environmental Knowledge, Meaning, Scale, pp. 203–217. Ashgate, Hampshire (2008)
42. Rambaldi, G., Chambers, R., McCall, M., Fox, J.: Practical ethics for PGIS practitioners, facilitators, technology intermediaries and researchers. Particip. Learn. Action **54**(1), 106–113 (2006)
43. Willis, N.: OpenStreetMap project imports US government maps (2011). https://www.linux.com/news/openstreetmap-project-imports-us-government-maps. Accessed 24 June 2018
44. Neteler, M., Bowman, M.H., Landa, M., Metz, M.: GRASS GIS: a multi-purpose open source GIS. Environ. Model Softw. **31**, 124–130 (2012)
45. Norheim-Hagtun, I., Meier, P.: Crowdsourcing for crisis mapping in Haiti. Innovations **5**(4), 81–89 (2010)

46. Seto, T.: Spatiotemporal transition of volunteered geographic information as a response to crisis: a case study of the crisis mapping project at the time of Great East Japan Earthquake. Papers and Proceedings of the Geographic Information Systems Association, vol. 20, B-2-4 (2011)

47. Kingsbury, P., Jones, J.P.: Walter Benjamin's Dionysian adventures on Google Earth. Geoforum **40**, 502–513 (2009)

48. Perkins, C., Dodge, M.: Satellite imagery and the spectacle of secret spaces. Geoforum **40**, 546–560 (2009)

49. Zook, M., Dodge, M., Aoyama, Y., Townsend, A.: New digital geographies: information, communication, and place. In: Brunn, S.D., Cutter, S.L., Harrington Jr., J.W. (eds.) Geography and Technology, pp. 155–176. Springer, Dordrecht (2004). https://doi.org/10.1007/978-1-4020-2353-8_7

50. Stefanidis, A., Crooks, A., Radzikowski, J.: Harvesting ambient geospatial information from social media feeds. GeoJournal **78**(2), 319–338 (2013)

51. Chaudhry, I.: # Hashtagging hate: using Twitter to track racism online. First Monday **20**(2) (2015). https://journals.uic.edu/ojs/index.php/fm/article/view/5450/4207. Accessed 23 June 2018

52. Yang, W., Mu, L.: GIS analysis of depression among Twitter users. Appl. Geogr. **60**, 217–223 (2015)

53. Boyd, D.M., Ellison, N.B.: Social network sites. J. Comput. Mediat. Commun. **13**, 210–230 (2007)

54. Debatin, B., Lovejoy, J.P., Horn, A.K., Hughes, B.N.: Facebook and online privacy: attitudes, behaviors, and unintended consequences. J. Comput. Mediat. Commun. **15**(1), 83–108 (2009)

55. Crampton, J.W.: Cartographic rationality and the politics of geosurveillance and security. Cartogr. GIS **30**(2), 135–148 (2003)

56. Dobson, J.E., Fisher, P.F.: Geoslavery. Technol. Soc. Mag. **22**(1), 47–52 (2003)

57. Klinkenberg, B.: Geospatial technologies and the geographies of hope and fear. Ann. Assoc. Am. Geogr. **97**(2), 350–360 (2007)

58. Propen, A.D.: Critical GPS: toward a new politics of location. ACME Int. J. Crit. Geogr. **4**(1), 131–144 (2005)

59. Dobson, J.E., Fisher, P.F.: The Panopticon's changing geography. Geogr. Rev. **97**(3), 307–323 (2007)

60. Koskela, H.: 'Cam Era'—the contemporary urban Panopticon. Surveill. Soc. **1**(3), 292–313 (2002)

61. Gross, R., Acquisti, A.: Information revelation and privacy in online social networks. In: Proceedings of the 2005 ACM Workshop on Privacy in the Electronic Society, pp. 71–80 (2005)

62. Jones, H., Soltren, J.H.: Facebook: threats to privacy. Proj. MAC MIT Proj. Math. Comput. **1**, 1–76 (2005)

63. Barnes, S.B.: A privacy paradox: social networking in the United States. First Monday **11**(9) (2006). http://dx.doi.org/10.5210/fm.v11i9.1394. Accessed 7 Aug 2018

64. McGrath, J.E.: Loving Big Brother: Performance, Privacy and Surveillance Space. Psychology Press, London (2004)

65. Ong, E.Y.L., et al.: Narcissism, extraversion and adolescents' self-presentation on Facebook. Pers. Individ. Differ. **50**(2), 180–185 (2011)

66. Shilton, K.: Four billion little brothers? Privacy, mobile phones, and ubiquitous data collection. Commun. ACM **52**(11), 48–53 (2009)

67. Rose-Redwood, R.S.: Governmentality, geography, and the geo-coded world. Prog. Hum. Geogr. **30**(4), 469–486 (2006)

68. Whitaker, R.: The End of Privacy: How Total Surveillance is Becoming a Reality. The New Press, New York (1999)
69. Kawaguchi, Y., Kawaguchi, K.: What does Google Street View bring about? -Privacy, discomfort and the problem of paradoxical others-. Contemp. Appl. Philos. **4**, 19–34 (2012)
70. Smith, P.K., Mahdavi, J., Carvalho, M., Fisher, S., Russell, S., Tippett, N.: Cyberbullying: its nature and impact in secondary school pupils. J. Child Psychol. Psychiatry **49**(4), 376–385 (2008)
71. Wolak, J., Finkelhor, D., Mitchell, K.J., Ybarra, M.L.: Online "predators" and their victims: myths, realities, and implications for prevention and treatment. Am. Psychol. **63**(2), 111–128 (2008)
72. Alexy, E.M., Burgess, A.W., Baker, T., Smoyak, S.A.: Perceptions of cyberstalking among college students. Brief Treat. Cris. Interv. **5**(3), 279–289 (2005)
73. Sheridan, L.P., Grant, T.: Is cyberstalking different? Psychol. Crime Law **13**(6), 627–640 (2007)
74. Kowalski, R.M., Limber, S.P.: Electronic bullying among middle school students. J. Adolesc. Health **41**, S22–S30 (2007)
75. Björkqvist, K., Lagerspetz, K.M., Kaukiainen, A.: Do girls manipulate and boys fight? Developmental trends in regard to direct and indirect aggression. Aggress. Behav. **18**(2), 117–127 (1992)
76. Ostrov, J.M., Keating, C.F.: Gender differences in preschool aggression during free play and structured interactions: an observational study. Soc. Dev. **13**(2), 255–277 (2004)
77. Card, N.A., Stucky, B.D., Sawalani, G.M., Little, T.D.: Direct and indirect aggression during childhood and adolescence: a meta-analytic review of gender differences, intercorrelations, and relations to maladjustment. Child Dev. **79**(5), 1185–1229 (2008)
78. Salmivalli, C., Kaukiainen, A.: "Female aggression" revisited: variable- and person-centered approaches to studying gender differences in different types of aggression. Aggress. Behav. **30**(2), 158–163 (2004)
79. Li, Q.: Cyberbullying in schools: a research of gender differences. Sch. Psychol. Int. **27**(2), 157–170 (2006)
80. Erdur-Baker, Ö.: Cyberbullying and its correlation to traditional bullying, gender and frequent and risky usage of internet-mediated communication tools. New Media Soc. **12**(1), 109–125 (2010)
81. Sui, D.Z.: Legal and ethical issues of using geospatial technologies in society. In: Nyerges, T.L., Couclelis, H., McMaster, R. (eds.) The Sage Handbook of GIS and Society, pp. 504–528. Sage, London (2011)
82. Awan, I.: Cyber-extremism: Isis and the power of social media. Soc **54**, 138–149 (2017)
83. Suzuki, K.: A newly emerging ethical problem in PGIS: ubiquitous atoque absconditus and casual offenders for pleasure. In: Proceedings of the 4th International Conference on GISTAM, pp. 22–27 (2018)
84. Vegh, S.: Classifying forms of online activism: the case of cyberprotests against the World Bank. In: McCaughey, M., Ayers, M.D. (eds.) Cyberactivism: Online Activism in Theory and Practice, pp. 71–95. Routledge, New York (2003)
85. Manion, M., Goodrum, A.: Terrorism or civil disobedience: toward a hacktivist ethic. ACM SIGCAS Comput. Soc. **30**(2), 14–19 (2000)
86. Suler, J.: The online disinhibition effect. CyberPsychology Behav. **7**(3), 321–326 (2004)
87. Lapidot-Lefler, N., Barak, A.: Effects of anonymity, invisibility, and lack of eye-contact on toxic online disinhibition. Comput. Hum. Behav. **28**, 434–443 (2012)
88. Dennis, K.: Keeping a close watch–the rise of self-surveillance and the threat of digital exposure. Sociol. Rev. **56**(3), 347–357 (2008)

89. Trottier, D.: Digital vigilantism as weaponisation of visibility. Philos. Technol. **30**(1), 55–72 (2017)
90. Herold, D.K.: Development of a civic society online? Internet vigilantism and state control in Chinese cyberspace. Asia J. Glob. Stud. **2**(1), 26–37 (2008)
91. Cheong, P.H., Gong, J.: Syber vigilantism, transmedia collective intelligence, and civic participation. Chin. J. Commun. **3**(4), 471–487 (2010)
92. Clarke, R.: Information technology and dataveillance. Commun. ACM **31**(5), 498–512 (1988)
93. Noeru: Kawasaki-jiken no yougisha taku ni iku haishin (2015). http://www.afreecatv.jp/33879426/v/59115. Accessed 4 Mar 2015
94. ABC News: YouTube star under fire for video of apparent suicide victim (2018). https://www.youtube.com/watch?v=WjNFGZLJLss. Accessed 19 Oct 2018
95. YouTube ni shogakusei wo nakasu douga wo toukou shita DQN wo Tsubusou @wiki (2012). https://www34.atwiki.jp/dqntokutei/. Accessed 19 Oct 2018
96. Jackson, W.H.: Chivalry in Twelfth-Century Germany: The Works of Hartmann von Aue, vol. 34. Boydell & Brewer Ltd., Cambridge (1994)
97. Lewis, C.T.: An Elementary Latin Dictionary. American Book Company, New York (1890)
98. Anonymous ed.: Ano AA doko? (2018). http://dokoaa.com/event/maturi.html#8. Accessed 19 Oct 2018
99. Westin, A.: Privacy and Freedom. Athenum, New York (1967)
100. Moor, J.H.: Towards a theory of privacy in the information age. Comput. Soc. **27**(3), 27–32 (1997)
101. Warren, S.D., Brandeis, L.D.: The right to privacy. Harv. Law Rev. **4**(5), 193–220 (1890)
102. Safari, B.A.: Intangible privacy rights: how Europe's GDPR will set a new global standard for personal data protection. Seton Hall Law Rev. **47**, 809–848 (2016)
103. noyb.eu: GDPR: noyb.eu filed four complaints over "forced consent" against Google, Instagram, WhatsApp and Facebook (2018). https://noyb.eu/wp-content/uploads/2018/05/pa_forcedconsent_en.pdf. Accessed 6 Aug 2018
104. Kramer, E., Ikeda, R.: Defining crime: signs of postmodern murder and the "Freeze" case of Yoshihiro Hattori. Am. J. Semiot. **17**(1), 19–84 (2001)
105. Hennigan, W.J.: Google pays Pennsylvania couple $1 in Street View lawsuit. Los Angeles Times, 2 December 2010. http://latimesblogs.latimes.com/technology/2010/12/google-lawsuit-street.html. Accessed 7 Aug 2018
106. Ling, Y.: Note: Google street view – privacy issues down the street, across the border, and over the seas. Boston University Journal of Science and Technology Law 2008 (2008). http://dx.doi.org/10.2139/ssrn.1608130. Accessed 7 Aug 2018
107. Federal Data Protection and Information Commissioner (2009). http://www.edoeb.admin.ch/dokumentation/00526/00529/00530/00534/index.html?lang=en. Accessed 12 Feb 2016
108. Federal Data Protection and Information Commissioner: Press release: verdict of the Federal Supreme Court in Google Street View case (2012). http://www.edoeb.admin.ch/datenschutz/00683/00690/00694/index.html?lang=en&download=NHzLpZeg7t,
lnp6I0NTU04212Z6ln1ad1IZn4Z2qZpnO2Yuq2Z6gpJCDdnx6hGym162epYbg2c_
JjKbNoKSn6A. Accessed 12 Feb 2016

Indexing Spelling Variants for Accurate Address Search

Konstantin Clemens[(✉)]

Technische Universität Berlin, Service-centric Networking, Berlin, Germany
konstantin.clemens@campus.tu-berlin.de

Abstract. In this paper, an alternative approach to handling erroneous user input is evaluated on the example of *address search* or *geocoding*. This process of resolving names of spatial entities like postal addresses or administrative areas into their whereabouts is error-prone for multiple reasons: Postal addresses are not structured in a canonical way and their format is only coherent within countries or regions; names of streets, districts, cities, regions, and even countries are often reused for their historical context; also human users often do not adhere to common address formats or do not spell out the names of the address entities completely nor correctly.

To handle the human-error, in this paper a log of address searches from real human users is used to construct a model of user behavior with regards to query format and spelling mistakes. This model is used to generate alternative spelling variants that are indexed in addition to the correctly spelled address entity names. Experiments show that extending the index of a geocoder this way yields higher recall and higher precision as compared to edit distances - the common approach to handling user queries with mistakes.

Keywords: Geocoding · Postal address search · Spelling variant · Spelling error · Document search

1 Introduction

Nowadays digital maps and digital processing of location information are widely used. Besides various applications for automated processing of location data, like [2,3,33], or [34], users rely on computers to navigate through an unknown area or to store, retrieve, and display location information. Withal, internally, computers reference locations through a coordinate system such as WGS84 latitude and longitude coordinates [23]. Human users, on the other hand, usually refer to locations by addresses or common names. The process of mapping such names or addresses to their location on a coordinate system is called *geocoding*.

There are two aspects to this error-prone process [11,15–17]: First, the geocoding system needs to parse the user query and derive the query intent, i.e., the system needs to understand which address entity the query refers to. Then, the system

© Springer Nature Switzerland AG 2019
L. Ragia et al. (Eds.): GISTAM 2018, CCIS 1061, pp. 73–87, 2019.
https://doi.org/10.1007/978-3-030-29948-4_4

needs to look up the coordinates of the entity the query was referring to and return it as a result. Already the first step is a non-trivial task, especially when considering the human factor: Some address elements are often misspelled or abbreviated by users in a non-standard way. Also, while postal addresses seem structured and as they would adhere to a well-defined format, [4] shows that each format only holds within a specific region. Considering addresses from all over the world, address formats often contradict to each other, so that there is no pattern that all queries would fit in. In addition to that, like with spelling errors, human users may not adhere to a format, leaving names of address elements out or specifying them in an unexpected order. Such incomplete or missorted queries are often ambiguous, as the same names are reused for different and oftentimes unrelated address elements. Various algorithms are employed to mitigate these issues. Even with the best algorithms at hand, however, a geocoding service can only be as good as the data it builds upon, as understanding the query intent is not leading to a good geocoding result if, e.g., there is no data to return.

Many online geocoding services like those offered by Google [18], Yandex [38], Yahoo! [37], HERE [19], or OpenStreetMap [25] are easily accessible by the end user. Because most of these systems are proprietary solutions, they neither reveal the data nor the algorithms used. This makes it hard to compare distinct aspects of such services. An exception to that is OpenStreetMap: The crowd-sourced data is publicly available for everyone to download. Open-source projects like Nominatim [24] provide geocoding services on top of that. In this paper, data from OpenStreetMap is used to create a geocoding service that can derive the user intent from queries, even if they contain spelling errors or are stated in non-standard formats. Nominatim - the reference geocoder for OpenStreetMap data - is used as one of the baselines, that the quality of service is compared with.

Geocoding systems try to return responses with one single correct results. However, sometimes no results might be retrieved for a query, while for ambiguous queries most geocoding systems yield responses with multiple results. Therefore, in this paper the *precision* of a geocoder is computed by the ratio of the number of responses containing the correct result over the number of responses containing any results at all, while as the *recall* is given by the ratio of the number of responses containing the correct result queried for to the total number of responses received and queried sent.

This paper is a continuation of the work made in [8]. A novel approach is suggested to increase both the recall and the precision of a geocoder. The idea is to make the system capable of supporting specific, most commonly made spelling errors. Usually, spelling variants in user queries are handled through allowing edit distances between tokens of the query and the address. That, however, inherently increases the ambiguity of queries and leads to lower precision and recall of the system: More responses contain results that queries did not refer to; also, while more responses contain the correct results, even more responses contain results at all. The suggested approach aims to avoid that by only allowing specific spelling variants that are made often while avoiding spelling variants that are not made at all or only made very rarely - edit distances lack this differentiation.

For that, from a log of real user queries the spelling mistakes observed most often are extracted. These spelling variants are indexed in addition to the correctly spelled address tokens. Variants of geocoding systems created this way are evaluated with regards to their precision and recall metrics, and compared to a similar system supporting edit distances, a baseline without spelling variants indexed or edit distances allowed, as well as Nominatim. In [5] and [6], similar measurements have shown that TF/IDF [31,32] or BM25f [29] based document search engines like Elasticsearch [13] handle incomplete or shuffled queries much better than Nominatim. This paper extends on that work. It adds to both the indexing mechanism proposed in [5] and [6] as well as the way the system performance is measured.

Work on comparing geocoding services has been undertaken in, e.g., [10,30, 39], or [12]. Mostly, such work focus on the recall aspect of a geocoder: Only how often a system can find the right result is compared. Also, other evaluations of geocoding systems treat every system as a black box. Thus, a system can be algorithmically strong but perform poorly in measurements because it is lacking data. Vice versa, a system can perform better than others just because of great data coverage, despite being algorithmically poor. In this paper, the algorithmic aspect is evaluated in isolation, as all systems are set up with the same data. Also, a different way of measuring the geocoders performance is proposed: The statistical model created based on real user queries is used to generate erroneous, user-like queries out of valid addresses. This allows measuring a system on a much greater number of queries, for which the correct response is known a priori.

Another approach to the geocoding problem is to find an address schema that is easy to use and standardized in a non-contradicting way. While current schemata of postal addresses are maintained by the UPU [35], approaches like [9,14,22,36], or [26] are suggesting standardized or entirely alternative address schemata. [7] shows that such address schemata are beneficial in some scenarios, though they are far from being adopted into everyday use.

In the next section, the approach and steps for setting up such geocoding systems are described in detail. Afterward, in Sect. 3 the undertaken measurements are explained. Next, in Sect. 4, the observed results are discussed and interpreted. Finally, in the last section, the conclusions are summarized and further work is discussed.

2   Setting up a Geocoder

The experiment is conducted on the OpenStreetMap data set for Europe. This data set is not collected with a specific application in mind. For many use cases, it needs to be preprocessed from its raw format before it can be consumed. As in [5] and [6], the steps for preprocessing OpenStreetMap data built into Nominatim have been used. Though a long-lasting task, reusing one same preprocessing mechanism ensures all geocoding systems are set up with exactly the same data, thereby enabling the comparability of the algorithmic part of those systems. Thus, first, Nominatim has been set up with OpenStreetMap data for Europe as

```
ID: 20
TEXT: Ernst Reuter Platz, 10587 Berlin
LATLON: 52.5127183,13.3217624
HOUSE NUMBERS:
- 7
    ID:200
    TEXT:Ernst Reuter Platz 7, 10587 Berlin
    LATLON: 52.5127183,13.3217624
- 9
    ID:201
    TEXT:Ernst Reuter Platz 9, 10587 Berlin
    LATLON: 52.5129359,13.3203243
```

Fig. 1. Example of a document without spelling variants, indexed in the geocoding system [8].

a baseline geocoding system. Internally Nominatim uses a PostGIS [27] enabled PostgreSQL [28] database. After preprocessing and loading the OpenStreetMap data, this database contains assembled addresses along with their parent-child relationships: A house number level address is the child of a street level address, which in turn is the child of a district level address, etc. This database is used to extract address documents that are indexed in Elasticsearch, as defined in [5] and [6]. Note that in this paper, only the geocoding of house number level addresses is evaluated. Therefore, though OpenStreetMap data also contains points of interests with house number level addresses, only their addresses but not their names have been indexed. Similarly, no parent level address elements, such as streets, postal code areas, cities, or districts have been indexed into Elasticsearch. All house number addresses with the same parent have been consolidated into one single document. Every house number has thereby been used as a key to specify the respective house number level address. Figure 1 shows an example document containing two house numbers 7 and 9, along with their WGS84 latitude and longitude coordinates and spelled-out addresses. The TEXT field of the document is the only one indexed; the TEXT fields mapped by the house numbers are only used to assemble a human-readable result.

Because Elasticsearch retrieves full documents, and because the indexed documents contain multiple house number addresses, a thin layer around Elasticsearch is needed to make sure only results with house numbers specified in queries are returned. That is a non-trivial task, as given a query, it is not known upfront which of the tokens is specifying the house number. Therefore, this layer has been implemented as follows: First, the query is split into tokens. Next, one token is assumed to be the house number; a query for documents is executed containing only the other tokens. This is repeated for each token, trying out every token as the house number. Because each time only one token is picked to specify the house number, this approach fails to support house numbers that are specified in multiple tokens. Nevertheless, it is good enough for the vast majority of cases. For every result document returned by Elasticsearch, the house number map is checked. If the token assumed to be the house number happens to be a key in that map, the value of that map is considered a match and the house number address is added to the result set. Finally, the result set is returned. As edit distances of query tokens can be specified in the query to Elasticsearch,

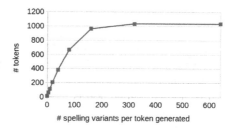

Fig. 2. Average number of tokens per document for various amounts of spelling variants [8].

this layer allows enabling edit distances easily: A parameter passed to the layer is forwarded to Elasticsearch, which then also returns documents with fuzzily matching tokens of address element names. Also note, that as house numbers are used as keys in the documents, neither edit distances nor spelling variants are supported on house numbers. That, however, is a natural limitation: If a query specifies a different house number than the one intended, especially if it is a house number that exists in the data, there is no way for a geocoding system to still match to the right house number value.

Having the baseline systems Nominatim and Elasticsearch supporting edit distances set up, the next step is to create a geocoding system that indexes spelling variants. For that, the spelling variants to be indexed need to be computed first. HERE Technologies, the company behind the HERE geocoding service [19], provided logs of real user queries issued against the various consumer offerings of the company, like their website or their applications for Symbian, Windows Phone, Android and iOS mobile phones. The log contained data from a whole year and included queries users have issued along with results users chose to click on. For this paper, a user click is considered the selection criterion of a result, linking input queries to their intent, i.e. the addresses users were querying for. Given such query and result pairs, first both were tokenized and the Levenshtein distance [21] from every query token to every result token was computed. With edit distances at hand, the Hungarian method [20] was used to align every query token to a result token. A threshold was used to ensure that if a user query contained a superfluous token that didn't match to any result token, it would remain unaligned. From these computations, several observations were extracted:

1. Some query tokens are superfluous as they do not match (close enough) to any result token. Such tokens are ignored.
2. As the result is a fully qualified address, result tokens have an address element type, such as city, street, house number, or country associated. Thus, for each query, the query format, i.e., which address elements were spelled out in what order, is known.

3. Some query tokens matched to result tokens had a Levenshtein distance greater than zero as they were misspelled. Thus, for each such spelling variant of a token, the specific spelling mistake made can be derived from the computation of the Levenshtein distance. For this paper, the following classes of spelling variants were considered:

  - **inserts:** Characters are appended after a tailing character, prepended before a leading character, or inserted between two characters, e.g., *s* is often inserted between the characters *s* and *e*, as apparently the double-s in *sse* sounds like a correct spelling for many users.
  - **deletes:** Characters are removed after a character that is left as the tailing one, before a character that is left as the leading one, or between two characters that are left next to each other, e.g., the characters *enue* after a tailing *v* are often deleted, as users often abbreviate *avenue* as *av.*
  - **replacements:** One character is replaced by a different character, e.g., *ß* is often replaced by an *s* in user queries so that *Straße* becomes *Strase* instead.
  - **swaps:** Two consecutive characters are swapped with each other, e.g., *ie* is often swapped into *ei*, as, to users, both spelling variants might sound similar.

Thus, from each query and result pair, the query format used as well as the set of spelling variations can be deduced. Doing so for the whole corpus of queries while counting the occurrences of each query format and each spelling variation results in two statistical models. One model can determine the possible spelling variations of a given token, each with their observed count or probability of occurrence. Out of a set of available address elements on a canonical address representation, the other model can select and order elements such that the resulting address representations correspond to query formats human users use, again, each with their observed count or probability of occurrence. Because the spelling mistakes made as well as the query formats used are Pareto distributed [1], the model contained a long tail of mistakes and formats used only very few times. To reduce that noise, the model was cleansed by stripping off the 25% of all observations from the long tail of rare spelling mistakes and query formats. In addition to that, all query formats that did not contain a house number were removed too, as the goal was to generate queries for addresses with house numbers. Because the log used is, unfortunately, proprietary, neither the log nor the trained models can be released with this publication. However, having a similar log of queries from another source enables the creation of a similar model.

Having the user model at hand, the spelling variants for indexing were derived as follows: Given a document to be indexed, its *TEXT* field was tokenized first. Next, for each token N most common spelling variants were fetched from the model and appended to the field. Thus, the field contained both the correctly spelled tokens as well as N spelling variants for each token. Every house number level address from Nominatim was extracted from the database, augmented with spelling variants and indexed in Elasticsearch. For N the values 10, 20, 40, 80, 160, 320, and 640 were chosen. A feature of this approach is that given a model that generates spelling variants the number of generatable spelling variations is limited. In most extreme cases, especially for short tokens, no spelling variant can be derived for a given token from the model at all. Figure 2 shows the resulting token counts of the *TEXT* field for every *N*. While for smaller amounts of generated the documents grow in their token count linearly to the number of generated tokens, the curve flattens at some point. There is only a minor increase between indexing 320 and 640 spelling variants, as with 320 spelling variants almost all observed variants have been generated already.

An interesting aspect of the described approach is that, besides lowercasing, no normalization mechanisms have been implemented. Users often choose to abbreviate common tokens like street types, or avoid choosing the proper diacritics and the model is capable of observing common replacements of, e.g., *Road* with *Rd*, or *Straße* with *Strasse* and generate corresponding spelling variants for indexing. Like with the index without spelling variants, house numbers are not modified in any way here.

In total, multiple geocoding systems were set up with the same address data indexed: Nominatim as the reference geocoder for OpenStreetMap data, Elasticsearch with documents containing aggregated house numbers along with a layer to support edit distances, and Elasticsearch with indexed spelling variants. While the edit distance was specified at query time by specifying a parameter to the layer wrapping Elasticsearch, for the various numbers of indexed spelling variants distinct Elasticsearch indices have been set up. As the same layer has been used for all Elasticsearch based indices, the setup supported the possibility to query an index with spelling variants while allowing an edit distance at the same time, thereby evaluating the effect of the combination of the two approaches.

3   Measuring the Performance

To evaluate the geocoding systems for recall - the ratio of responses containing the right result to all results, and precision - the ratio of responses containing the right result to responses containing any result, 50000 existing addresses have been sampled from the data used in these systems. Using generative user models trained on a dedicated subset of the data, first, for each address a query format has been chosen so that the distribution of the query formats chosen corresponded to the observed distribution of the query formats preferred by users: Commonly used query formats were selected more often than rarely used ones. Next, for each query one to five query tokens have been chosen to be replaced with a spelling variant. Again, spelling variants picked were distributed in the same way the spelling variants of human users were distributed: Frequent spelling mistakes were present in the test set often, while rare spelling variants were present rarely. This way six query sets with 50000 queries each have been generated. All queries were in a format that users have used to query. One query set contained all tokens in their original form, while the other sets had between one and five query tokens replaced with a spelling variant. Note that a query had not always the desired number of spelling variants: The token to be

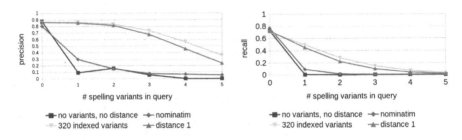

Fig. 3. Precision (left) and recall (right) of select geocoding systems with different approaches.

replaced with a spelling variant was chosen at random. For some tokens, as discussed, no spelling variant could be generated by the model. These tokens were left unchanged, making the query contain fewer spelling variants than anticipated in some cases. Also, sometimes the house number token was chosen to be replaced. Given the setup of the documents in the indices, where house numbers are used as keys in a map, such queries had no chance of being served properly. This, however, does not pollute measurement results, as it equally applies to all systems evaluated. Because generated queries and indexed addresses originate from the same Nominatim database, both share the same unique address identifier. Therefore, inspecting the result set of a response for the correct result to a query - the address the query has been generated from - is a simple task.

Each test set was issued against indices with 10, 20, 40, 80, 160, 320, and 640 indexed spelling variants, against the index with no spelling variants that allowed edit distances of 1 and 2, and against the two baselines: An index with neither spelling variants indexed nor edit distances allowed, as well as Nominatim. Additionally, each query set was issued against the combination of the two approaches: Indices with spelling variants indexed were queried with edit distances allowed. For every query set, responses were inspected to contain any results and to contain the right result using the address identifier. From all requests, for every query set and geocoding system, the precision and recall ratios were computed.

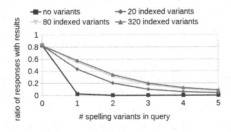

Fig. 4. Ratio of responses with any results of systems with various amount of indexed spalling variant.

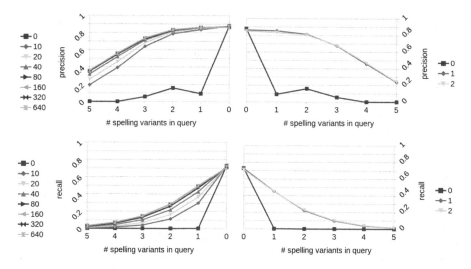

Fig. 5. Detailed overview of precision (top) and recall (bottom) of geocoding systems with spelling variants (left, each chart denotes the number of spelling variants indexed) and edit distances (right, each chart denotes an allowed edit distance).

Fact is, knowing the distributions of spelling variants, and how many of them were indexed, it is possible to calculate how many responses will include the expected result up front, without any measurement: The portion of spelling variants indexed is exactly the portion of spelling variants in queries that a geocoder will be able to serve with the correct result. However, while this way we could compute the expected recall, there is no way to calculate the increase of ambiguity of documents when indexing spelling variants - there is no way to assess up front how many queries will contain tokens that are spelling variants of different documents. Thus, we cannot compute how many queries will contain any result at all, and hence to compute the value for precision. The measurements undertaken for this paper allow observing the development of precision and recall metrics while the number of indexed spelling variants or the number of supported edit distances is increased step by step, and compare the two approaches with themselves and with sane baselines.

Another interesting question to investigate was: How much does the model itself alter over time? Users might change the type of mistakes they are making, thereby rendering some of the spelling variants indexed in a geocoding system obsolete: While they used to be common and important to be indexed, they might no longer be of relevance for current queries. To observe that, a dedicated experiment has been conducted as follows: Two models were trained each on another half of the query log provided by HERE Technologies. Thus, both models were built from independent observations - one in the first half of the year, and the other in the second half. Both models have been used to generate query sets with zero to five spelling variants similarly to the previous experiments. Also, both models have been used to build two geocoders indexing 320 spelling variants each. Each query set was measured for precision and recall against the index generated with the same model used to generate the query set, as well as with the index generated by the other model. This way, one could observe how the index performance deteriorates when it is measured against a query set mimicking user behavior from another point in time.

4 Results

Figure 3 shows an overview of precision and recall of select systems with different approaches. The red chart of Nominatim shows that for queries with none or one spelling mistake it performs better compared to the blue chart of Elasticsearch with neither spelling variants indexed nor edit distances allowed. That holds for both precision and recall and is likely due to the normalization mechanisms that are built into Nominatim, but missing in Elasticsearch: Some portion of commonly made spelling variants can be handled through normalization. For no spelling mistakes, both - Nominatim and Elasticsearch without spelling variants or edit distances show slightly higher recall compared to the yellow and green charts plotting the performance of the index with 320 spelling variants per token indexed, and the recall of enabling the edit distance of one, respectively. The higher ambiguity of the data in the latter two systems is the likely reason for this. As expected, the more spelling variants are present in queries the more precision and recall drop for all systems. While Nominatim and Elasticsearch without spelling variants or edit distances can barely handle queries with one mistake, both mechanisms to handle such queries function as expected: Their precision and recall drop much slower than that of the former two systems. Without exception, the index with 320 spelling variants per token indexed outperforms the index allowing an edit distance of one.

Note that one aspect of the continuously reduced recall is driven by spelling variants of house numbers. While none of the systems are laid out to handle spelling variants in house numbers, the generated queries allowed for such scenarios. The more spelling variants were generated in a query, the greater the chance for a modified house number.

A good overview of the ambiguity of the data is presented by Fig. 4 plotting the ratio of responses with any result for indices with various amount of spelling variants. Clearly, indexing no variants yields almost no results for queries with

Table 1. Configurations yielding best precision.

typos in query	0	1	2	3	4	5
variants indexed	0	160	640	160	640	160
edit distance	0	0	0	0	1	1
precision	87.88%	86.06%	83.41%	73.63%	56.62%	36.34%
recall	72.90%	49.37%	28.36%	14.37%	7.45%	3.33%

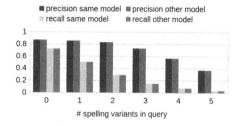

Fig. 6. Precision and recall of geocoders and test queries generated by models trained on one same or on two different time spans. (Color figure online)

typos: The misspelled tokens are simply not retrieved from the index. Also, as expected, adding 20 most-made spalling variants to the index vastly increases the number of retrieved results. However, because most-common spelling variants are added to the index first, the number of additional results retrieved grows slower, with further spelling variants added to the index: The increase in the number of results when increasing the number of indexed spelling variants from 20 to 80 is much larger compared to increasing that amount from 80 to 320. The most plausible, commonly made spelling variants have been added already, and the index is extended with only rare misspellings from the long-tail of their distribution. This exact property of the suggested approach results in better precision and recall as compared to allowing edit distances: Indexing more spelling variants does not result in more ambiguity between the documents at some point. Another thing observable on the graph is how more spelling variants in a query yield fewer results: Fewer tokens from the query are discovered in the indexed data resulting in fewer results at all.

The detailed experiment results are denoted in Fig. 5. The two top graphs show the precision of the indexes with various amount of spelling variants indexed on the left, and various allowed edit distances on the right. The bottom two graphs show the recall of the systems under test in a similar way. Unsurprisingly, the more spelling errors a query contains, the lower precision and recall is yielded by all systems consistently. Interestingly, increasing the allowed edit distance from one to two does not improve the precision nor the recall significantly on any test set. Similarly, at some point, the number of generatable spelling variants for a token is maxed out. Hence the increase in precision and recall of the

Table 2. Configurations yielding best recall.

typos in query	0	1	2	3	4	5
variants indexed	10	640	320	640	160	160
edit distance	1	1	1	1	1	1
precision	87.42%	85.42%	82.80%	73.23%	56.60%	36.34%
recall	73.85%	51.05%	29.36%	15.39%	7.46%	3.33%

indices with spelling variants indexed is not constant and vanishes at some point above 80 spelling variants indexed so that so that there is almost no difference between, e.g., indexing 320 or 640 spelling variants. The graphs are laid out so that a comparison between the two approaches can be made visually: Systems indexing spelling variants constantly yield a higher precision and a higher recall as compared to the systems allowing edit distances.

Note that the blue charts denoting the precision and recall of the index containing zero spelling variants on the left are equal to the blue charts denoting the performance of the system allowing no edit distances. That is because the index allowing zero spelling variants is conceptually equivalent to the index allowing an edit distance of zero - it is the plain Elasticsearch document search engine with no normalization that we use as a baseline besides Nominatim.

Tables 1 and 2 show the combinations of indexed spelling variants and allowed edit distances that lead to best precision and best recall respectively. For precision, the best results are retrieved with 160 to 640 spelling variants indexed. Only for queries with four or five spelling mistakes, allowing an edit distance increases precision. For better recall, an allowed edit distance of one helps consistently. Even with no spelling variants in the query, the recall is increased when allowing an edit distance of one and indexing 10 spelling variants: More correct results are retrieved along with more results overall so that this configuration is not the best one for precision.

Finally, Fig. 6 illustrates the deterioration of a model with regards to time. Four bars plot precision and recall metrics for both scenarios measured: The performance of a geocoding system powered by spelling variants generated through the same model used to generate the query sets, as well as the performance of the set-up where different models trained on data from different time spans were used for the index and for the query set. As visible on the graph, for any amount of spelling mistakes in the query, the precision and recall metrics are almost not affected by the model used to generate the queries and the model used to index the spelling variants.

5   Conclusion

As already observed in previous papers, here too, Nominatim does not handle human-like queries with mistakes well. A plain document search engine like Elastic search performs even worse as it lacks the necessary normalization mechanisms. Using a statistical model to derive and index common spelling variants, however, has proven to be a viable approach to serve queries with spelling errors. Compared to allowing edit distances, it consistently yields a higher precision and a higher recall. Interestingly, this approach implicitly incorporates any standardization logic necessary: Exactly those abbreviations or misspelled diacritics that are commonly made are indexed as spelling variants. The experiment also suggests that indexing more plausible spelling variants is generally better: No number of indexed spelling variants smaller than the maximum turned out to be the optimum beyond which performance of the index would degrade. Also, while indexed spelling variants outperform edit distances on all query sets, a combination of the two showed slightly better results, especially for queries with many spelling mistakes.

Another observation made was that models generated in such way seem to generalize fine over time: A model of user-made spelling variants based on the set of query logs from one half of the year is performing as well with queries generated by a model from the same time, as with queries generated from a model trained on the other half of the year. Possibly, however, with larger time spans or gaps between them, an impact on the performance of a geocoder becomes apparent.

Going forward, it is worth investigating how spelling variants can be indexed without obtaining a statistical user model first. In this paper, user clicks were used to learn how often and which typos are made. Users, however, can only click on the results they receive. Thus, a query token may be spelled so significantly different, that the system will not present the proper result to the user. Even if that spelling variant would be common, without a result to click on, no model could learn that spelling variant so that it can be indexed. Further, the set of supported spelling variants might be defined more precisely. The model could learn a larger context of an edit, like, e.g., four or more characters that surround an observed edit, as opposed to two characters only. Pursuing this idea to its full extent, a model could learn specific spelling variants for specific tokens instead of extrapolating observations from one token to another.

References

1. Arnold, B.C.: Pareto distribution. Wiley, Hoboken (2015)
2. Borkar, V., Deshmukh, K., Sarawagi, S.: Automatically extracting structure from free text addresses. IEEE Data Eng. Bull. **23**(4), 27–32 (2000)
3. Can, L., Qian, Z., Xiaofeng, M., Wenyin, L.: Postal address detection from web documents. In: 2005 International Workshop on Challenges in Web Information Retrieval and Integration, (WIRI2005), pp. 40–45. IEEE (2005)
4. Clemens, K.: Automated processing of postal addresses. In: GEOProcessing 2013: The Fifth International Conference on Advanced Geograhic Information Systems, Applications, and Services, pp. 155–160 (2013)
5. Clemens, K.: Geocoding with openstreetmap data. In: GEOProcessing 2015: The Seventh International Conference on Advanced Geograhic Information Systems, Applications, and Services, p. 10 (2015)
6. Clemens, K.: Qualitative comparison of geocoding systems using openstreetmap data. Int. J. Adv. Softw. **8**(3 & 4), 377 (2015)
7. Clemens, K.: Comparative evaluation of alternative addressing schemes. In: GEOProcessing 2016: The Eighth International Conference on Advanced Geograhic Information Systems, Applications, and Services, p. 118 (2016)
8. Clemens, K.: Enhanced address search with spelling variants. In: Proceedings of the 4th International Conference on Geographical Information Systems Theory, Applications and Management - Volume 1: GISTAM, pp. 28–35. INSTICC, SciTePress, Setúbal (2018). https://doi.org/10.5220/0006646100280035
9. Coetzee, S., et al.: Towards an international address standard. In: 10th International Conference for Spatial Data Infrastructure (2008)

10. Davis, C., Fonseca, F.: Assessing the certainty of locations produced by an address geocoding system. Geoinformatica **11**(1), 103–129 (2007)
11. Drummond, W.: Address matching: Gis technology for mapping human activity patterns. J. Am. Plan. Assoc. **61**(2), 240–251 (1995)
12. Duncan, D.T., Castro, M.C., Blossom, J.C., Bennett, G.G., Gortmaker, S.L.: Evaluation of the positional difference between two common geocoding methods. Geospatial Health **5**(2), 265–273 (2011)
13. Elastic: Elasticsearch, June 2017. https://www.elastic.co/products/elasticsearch
14. Fang, L., Yu, Z., Zhao, X.: The design of a unified addressing schema and the matching mode of china. In: 2010 Geoscience and Remote Sensing Symposium (IGARSS). IEEE (2010)
15. Fitzke, J., Atkinson, R.: OGC best practices document: Gazetteer service-application profile of the web feature service implementation specification-0.9. 3. Open Geospatial Consortium (2006)
16. Ge, X., et al.: Address geocoding, 23 August 2005
17. Goldberg, D., Wilson, J., Knoblock, C.: From text to geographic coordinates: the current state of geocoding. URISA J. **19**(1), 33–46 (2007)
18. Google: Geocoding API, June 2017. https://developers.google.com/maps/documentation/geocoding/
19. HERE: Geocoder API Developer's Guide, June 2017. https://developer.here.com/rest-apis/documentation/geocoder/
20. Kuhn, H.W.: The hungarian method for the assignment problem. Naval Res. logistics Q. **2**(1–2), 83–97 (1955)
21. Levenshtein, V.I.: Binary codes capable of correcting deletions, insertions, and reversals. Sov. Phys. Dokl. **10**, 707–710 (1966)
22. Mayrhofer, A., Spanring, C.: A uniform resource identifier for geographic locations ('geo'uri). Technical report, RFC 5870, June 2010
23. National Imagery and Mapping Agency: Department of Defense, World Geodetic System 1984, Its Definition and Relationships with Local Geodetic Systems. In: Technical Report 8350.2 Third Edition (2004)
24. OpenStreetMap Foundation: Nomatim, June 2017. http://nominatim.openstreetmap.org
25. OpenStreetMap Foundation: OpenStreetMap, June 2017. http://wiki.openstreetmap.org
26. geo poet, Jun 2017. http://geo-poet.appspot.com/
27. PostGIS, June 2017. http://postgis.net/
28. PostgreSQL, June 2017. http://www.postgresql.org/
29. Robertson, S., Zaragoza, H., Taylor, M.: Simple BM25 extension to multiple weighted fields. In: Proceedings of the Thirteenth ACM International Conference on Information and Knowledge Management, pp. 42–49. ACM (2004)
30. Roongpiboonsopit, D., Karimi, H.A.: Comparative evaluation and analysis of online geocoding services. Int. J. Geogr. Inf. Sci. **24**(7), 1081–1100 (2010)
31. Salton, G., Yang, C.S.: On the specification of term values in automatic indexing. J. Documentation **29**(4), 351–372 (1973)
32. Salton, G., Yang, C.S., Yu, C.T.: A theory of term importance in automatic text analysis. J. Am. Soc. Inf. Sci. **26**(1), 33–44 (1975)
33. Sengar, V., Joshi, T., Joy, J., Prakash, S., Toyama, K.: Robust location search from text queries. In: Proceedings of the 15th Annual ACM International Symposium on Advances in Geographic Information Systems, p. 24. ACM (2007)
34. Srihari, S.: Recognition of handwritten and machine-printed text for postal address interpretation. Pattern Recogn. Lett. **14**(4), 291–302 (1993)

35. Universal Postal Union, June 2017. http://www.upu.int
36. what3words: what3words, June 2017. https://map.what3words.com/
37. Yahoo!: BOSS Geo Services, June 2017. https://developer.yahoo.com/boss/geo/
38. Yandex: Yandex. Maps API Geocoder, June 2017. https://tech.yandex.com/maps/geocoder/
39. Yang, D.H., Bilaver, L.M., Hayes, O., Goerge, R.: Improving geocoding practices: evaluation of geocoding tools. J. Med. Syst. **28**(4), 361–370 (2004)

Geovisualization Technologies Applied to Statistical Characterization of the Iberian Peninsula Rainfall Using the Global Climate Monitor Web Viewer

Mónica Aguilar-Alba[✉], Juan Mariano Camarillo Naranjo,
and Juan Antonio Alfonso Gutiérrez

Departamento de Geografía, Universidad de Sevilla, 41004 Seville, Spain
{malba, jmcamarillo}@us.es

Abstract. The high variability that characterises most of the Iberian Peninsula climate types has important consequences in terms of water resources availability and climatic risks. The concern about the high spatial-temporal variability of precipitation in Mediterranean environments is accentuated in the Iberian Peninsula by its complex physiographic characteristics. However, despite its importance in water resources management it has been scarcely addressed from a spatial perspective. In general, precipitation is characterized by positive asymmetric frequency distributions at different time scales, so conventional statistical measures lose representativeness. For this reason, a battery of robust and non-robust statistics of little used in the characterization of precipitation has been calculated and evaluated quantitatively. The fact that the median normally has a lower value than the mean determines that the non-robust statistical measures could overestimate the input water. In Spain, hydrological modelling for water management and planning is based on mean precipitation values. This study aims to quantify different statistical measurements using long precipitation series provided by The Global Climate Monitor (GCM) (http://www.globalclimatemonitor.org). This Web-based GIS for geovisualization is one of the few examples that exist with regard to the diffusion of climatic data. The statistical characterisation of the Iberian Peninsula precipitation show important differences that might have significant consequences in the estimation and management of water resources. This study has been completely developed using Open Source technologies and has implied the design and management of a spatial database. The results are mapped using GIS and are incorporated into a web geoviewer (https://qgiscloud.com/Juan_Antonio_Geo/expo) in order to facilitate access to them.

Keywords: Geoviewers · Geospatial information and technologies · Precipitation · Spatio-temporal database management · Iberian Peninsula

1 Introduction

Current climate research benefits from the existence of large and global climate databases produced by various international organizations. The common denominator is the availability and accessibility under the 'open data' paradigm. Very often, these

© Springer Nature Switzerland AG 2019
L. Ragia et al. (Eds.): GISTAM 2018, CCIS 1061, pp. 88–107, 2019.
https://doi.org/10.1007/978-3-030-29948-4_5

new datasets cover the entire earth with a regular spatial distribution (normally gridded), with a longer and more homogeneous time span and are built under more-robust procedures [1, 3]. Given the advantages of these databases, our study aims to achieve a spatial characterization of precipitation in the Iberian Peninsula based on long historical series. The importance of this type of studies is determined by the importance of precipitation in water resources management in Mediterranean environments [1].

Many varied sources of information are at the basis of global climatic datasets are accessible on web portals which are normally distributed under *open-database licenses.* This has favored the increasingly widespread use of global data by scientists, and the emergence of countless references from studies based on these data [10, 17, 25]. However, the complexity of the very technical formats of distribution (NetCDF or huge plain text files with millions of records) limits these datasets to a very small number of users, almost exclusively scientists. For non-expert users, it is important to offer new environmental tools in an open and transparent framework so stakeholders, users, policy makers, scientists and regulators can make real use of them [4, 18].

These open data and open knowledge paradigm also referred by some as "the fourth paradigm" [9] responds, in relation to climate data, to the double challenge that climate science is currently facing of guaranteeing the availability of data to permit more exploration and research and to reach citizens. This leads directly to the use of open source programs supported by an extensive worldwide community of users that provide tested evidence in very stressful applications. Such a large and proven implementation in results and experience has also motivated the use of these open technologies in our research. Particularly, the Climate Research Group of the University of Seville has a remarkable experience with PostgreSQL/PostGIS being the core of the data management and research tools.

It is worth noting that geo viewers and have become one of the most useful web services for many applications and for decision support [25]. Also, web-mapping has become an effective tool for public information access and for climate monitoring, especially considering the ongoing advances in web GIS and geovisualization technologies.

Despite the fact that there are some very specialized geoviewers to access and download climatic data (for example, the *Global Climate Station Summary* by *NCDC-NOAA),* in most cases they offer poor geo-displays capabilities or too general information (European climate assessment). The Global Climate Monitor (GCM) system is built on a spatio-temporal database management that applies geospatial information technologies and web viewers to make this information accessible for many climatic applications. In this work we focus on showing the possibilities of the GCM in climatic research analyzing the spatial distribution of precipitation in the Iberian Peninsula.

Understanding the spatial distribution of rainfall is important for natural resources management in the Iberian Peninsula due to its extremely variable spatio-temporal distribution. With the exception of the northern mountain range, the Iberian Peninsula is included in the Mediterranean climate domain [8, 22] characterized the inter-annual irregularity of precipitation [11]. There are marked fluctuations from rainy years to periods of drought together with an irregular intra-annual regime with minimum values of rainfall during the summer months [12]. These makes decision-making in water management complicated requiring accurate scientific information to support technical decisions in the water planning process which directly affects the environment and society [2, 19].

Much research has been devoted to study rainfall regime in the Iberian Peninsula from the analysis of the annual, seasonal or monthly volumes to synoptic situations mainly linked to atmospheric circulation patterns and weather types [5, 7, 15, 20, 24, 26]. Nevertheless, there are not many studies devoted to the spatial rainfall distribution analysis. The present study introduces a complementary aspect to these types of works calculating a set of robust and non-robust statistics for the characterization of precipitation. Robust statistical measures are not commonly used when characterizing neither precipitation nor other water resources variables despite the positive skewness of their frequency distributions. This is the reason why we aim to study the differences between both types of measurements in order to evaluate the possible bias when estimating precipitation volumes.

This work has involved the implementation of a spatial database that handles a large amount of information allowing the analysis and processing of spatial data. Results can be viewed using GIS technologies and this information is disseminated and made accessible to final building up a new geoviewer (https://qgiscloud.com/Juan_Antonio_Geo/expo).

2 Study Area and Data

The Iberian Peninsula is located in the southwestern end of Europe, next to North Africa and surrounded by the Mediterranean Sea to the East and by the Atlantic Ocean to the West. Due to this transition position between the middle and the subtropical latitudes together with a complex orography determine complex patterns of spatio-temporal of most climatic variables [11].

The GCM currently displayed corresponds to the CRU TS3.21 version of the Climate Research Unit (University of East Anglia) database [6] that provides data at a spatial resolution of half a degree in latitude and longitude, spanning from January 1901 to December 2012 on a monthly basis. From January 2013, the datasets that feed the system are the GHCN-CAMS temperature dataset (Global Historical Climatology Network-Climate Anomaly Monitoring System) and the GPCC First Guess precipitation dataset (Global Precipitation Climatology Centre) from the Deutscher Wetterdienst in Germany [14].

The data that are currently offered in the display come from three main datasets: the CRU TS3.21, the GHCN-CAMS, and the GPCC first guess monthly product. The basic features of these products are shown in Table 1. The CRU TS3.21 dataset is a high-resolution grid product that can be obtained through the British Atmospheric Data Centre website or through the Climatic Research Unit website. It has been subjected to various quality controls and homogenization processes [23]. It is important to note that this database is also offered in the data section of their website and in their global geovisualization service [14].

Table 1. Basic features of the datasets included in the global climate monitor.

Dataset	CRU TS3.21	GHCN-CAMS	GPCC
Spatial resolution	$0.5° \times 0.5°$	$0.5° \times 0.5°$	$1° \times 1°$
Time span	January 1901 to December 2012	January 1948 to present expired month	August 2004 to present expired month
Spatial reference system	WGS84	WGS84	WGS84
Format	NetCDF	Grib	NetCDF
Variables	Total precipitation Average temperature	Average temperature	Total precipitation

Apart from this organization, CRU TS is one of the most widely used global climate databases for research purposes. The GHCN and GPCC datasets are used to update monthly data. The data used for this study are historical series of monthly precipitation from January 1901 to December 2016 that covers the entire Iberian Peninsula. Visually, they are represented in the form of a grid, in an area of $0.5° \times 0.5°$ latitude-longitude (see Fig. 1). Therefore, the total volume of information takes into account the total months, years and cells that compose the spatial extent. The amount of information of more than 400,000 records with spatial connection requires the use of a database management system. Given the nature of the precipitation data grid, which is not graphically adjusted to the actual physical limits of the Iberian Peninsula, it is necessary to adapt the rainfall information to the limits of the field of study. For this purpose a vector layer containing the physical boundaries of the different territorial units that make up the study area (Spain and Portugal) was also used. The 1: 1,000,000 territorial statistical units (NUTS) of the European territory for the year 2013 data have been obtained from the European Statistical Office (Eurostat, http://ec.europa.eu/eurostat/). The Coordinate Reference System through which this information is distributed is ETRS89 (EPSG: 4258), which is an inconvenience when combining this with the rainfall data projected in WGS84 Spatial Reference System used in the GCM. Therefore, through a geoprocessing of coordinate transformation both systems were squared.

The assembly of the available open-source technologies used in this study is shown in Table 2. The use of spatial data server, map server, web application server and web viewers allow scientists to undertake these types of macro-projects based on the use of Big Data information.

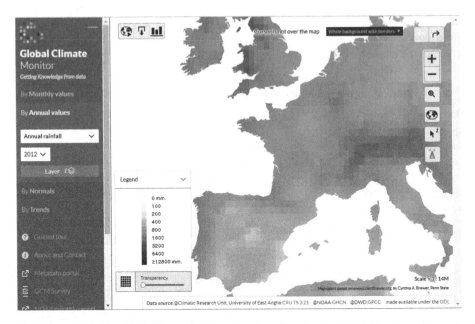

Fig. 1. View of the Global Climate Monitor interface (http://www.globalclimatemonitor.org).

Particularly noteworthy is the role of the spatial data server PostGIS in handling spatial and alphanumeric information and charts based on a relational system PostgreSQL. The data are natively encoded in said system providing a high performance and allowing the use of any analytical functions required; but the most importantly reason for using PostgreSQL/PostGIS is its geographic relational database that make possible to carry out geoprocessing without having to leave the processing core. This allows the regionalization of data in a quasi-automatic way for any territorial scales of interest based on SQL language.

Table 2. Open-source technologies used in this study.

Software	Application
PostgreSQL/PostGIS	Spatial database management system
PgAdmin	Administration and development platform for PostgreSQL
QGIS	GIS desktop
QGIS Cloud	Geoviewer web

There are other tools in this field of work that are more or less equivalent as the ones used in our study such as SciDB database, that manages multidimensional cubes and are especially suitable for satellite images processing, or some scientific libraries found in R or Python. Nevertheless, these technologies present a different approach. While it is true that for raster spatial applications time series data approach is the ideal environment to work with, it can be argued that, concerning the structure of the data, the same effect can be obtained with a more conventional relational approach. Such is again the case of R and Python programming language that cover the same scientific needs with different technologies. In any case, a geographic relational database can also replace these technologies in many scenarios like the one presented in this work. Another advantage is that PostgreSQL/PostGIS can be directly coupled to R (http://www.joeconway.com/doc/doc.html) thus obtaining a geostatistical database and enlarging the possibilities of use and applications.

Through QGIS Cloud a web geoviewer was developed as an extension of the QGIS. The resulting maps of this study are represented in a geoviewer (https://qgiscloud.com/Juan_Antonio_Geo/expo). Open source systems fit the main aim of the Global Climate Monitor project: rapid and friendly geovisualization of global climatic dataset and indicators to experts and non-experts users instead of data analytics.

3   Methodology

To carry out this study aiming the spatial characterization of precipitation in the Iberian Peninsula different challenges has been achieved. The main methodological steps based on a series of processes and procedures are presented in Fig. 2. This requires a previous design to establish a logical structure and requirements of the data in order to be able to model the information and analysis.

First, the data were downloaded from the GCM in csv format and converted into a database to be modelled. The proposed theoretical model defines the conceptual and relational organization which generates the physical database itself. The conceptual model is simple and consists of several tables: the meteorological stations table, the monthly precipitation table values and the table with the geographical limits of the Iberian Peninsula. Then, a series of queries are performed in SQL language on the monthly precipitation series of data in order to get the seasonal and annual values as well as the statistical measures calculated from them.

The contribution of using GIS and geospatial databases is that both allow massive geospatial time and space analysis and geovisualization. The aim is to get a series of statistics that will help us characterize the spatial distribution of the precipitation gridded series. The analysis of variability, dispersion, maximum and minimum and frequency histograms of the precipitation series in the Iberian Peninsula determine the need of using adequate statistical measures to fulfil our goals.

In the Mediterranean climate precipitation is extremely variable so the frequency distributions tend to be asymmetrical. Figure 3 shows the histograms of four main cities of the Iberian Peninsula placed in different climatic regions. Lisbon, Seville and Valencia correspond to a Köppen classification type of Csa climate (Mediterranean climate) whereas Santander located in the northern coast corresponds to a Cfb type (Temperate climate).

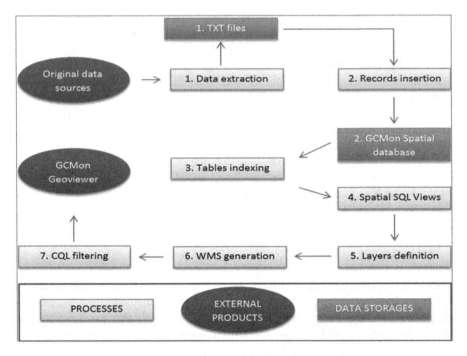

Fig. 2. Methodology flow chart.

The four annual precipitation histograms are positively skewed with less frequent high extreme values but very important due to the fact that often these values are related with flooding episodes. This extreme variability that characterizes most of the Iberian Peninsula climate has important consequences in terms of water resources availability and climatic risks. This is why it becomes so important to have an accurate characterization of the spatial precipitation distribution for environmental and water management. The fact that the median has always a lower value than the mean determines that the non-robust statistical measures might not be representative for the description and quantification of water availability.

It should be notice that the annual time scale does not show the inter-annual variability of these climates with dry and hot summers and rain concentrated mainly in winter and autumn. This fact is even more pronounced when considering seasonal or monthly series showing more marked positive skewed histograms. Therefore, the task of improving the general estimation of precipitation at different time scales with spatial continuity is a relevant issue for the whole of the Iberian Peninsula.

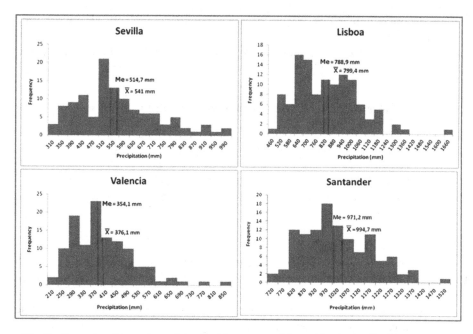

Fig. 3. Example of annual precipitation histograms of four cities in the Iberian Peninsula.

Typically, non-robust statistics (mean, standard deviation and variation coefficient) are commonly used for climate characterization. But these measures can be affected by extreme values that may detract their statistical representativeness. For this reason we decided to incorporate robust statistics to eliminate this effect and to be able to assess the differences between them (robust and non-robust) in both absolute and relative statistics (see Table 3).

Table 3. Statistical measures calculated for each series.

Measures	Central tendency	Absolute dispersion	Relative dispersion
Non robust	Mean ($\overline{X}$)	Standard Deviation (s)	Coefficient of Variation (CV)
Robust	Median (Me)	Interquartile Range (IRQ)	Interquartile Coefficient of Variation (ICV)
Difference	($\overline{X}$ – Me)	(s – IRQ)	

Database system allows the analysis of the historical precipitation series considering all monthly values recorded to calculate from the seasonal and annual series. It is possible then to study the statistical behavior and variability of precipitation at different time scales.

It is important to note that, despite the relative simplicity of these calculations, they offer many results due to the powerful data management capabilities of the spatial databases. Table 4 shows the statistical measures calculated at different time scales (annual, seasonal and monthly) for each precipitation series.

Table 4. Statistical measures calculated at different time scales.

	Measures		
	Central tendency	Absolute dispersion	Relative dispersion
Annual	3	3	2
Seasonal	12	12	8
Monthly	36	36	24
Total	51	51	34

The result has provided a total of 136 outcomes, which were viewed using GIS with the corresponding cartographies for the entire Iberian Peninsula. These results are obtained for each cell of the grid and incorporated into the open source Geographic Information System QGIS. Once incorporated into a GIS system web cartographic visualization can be represented and directly viewed. This is very useful tool when carrying out multitude of analysis processes or simply performing a cartographic representation of the results. Finally, the results are included in a new web geoviewer (https://qgiscloud.com/Juan_Antonio_Geo/expo).

4   Results

The management of large volumes of precipitation information through spatial data-bases has made possible to obtain products of relevant climatic interest related to the spatial estimation of precipitation. The main results presented in this paper are a wide range of statistical measurements of central tendency and measures of variability for the annual and seasonal series. The comparison and quantification of non-robust and robust statistical measures are spatially represented in GIS maps. Relative dispersion statistical measures are also calculated to diminish the effect of the very different rainfall amounts recorded in the Iberian Peninsula.

The cartographic representation of the central tendency measures show the three large homogeneous climatic zones traditionally described for the Iberian Peninsula this geographic coherence (see Fig. 4).

The first, called the Atlantic or humid region, corresponds to the Oceanic climates, which extend over most of the North coast from Galician to the Pyrenees; the second, corresponding to semi-arid or sub-desert region, occupies the southeast of the peninsula around the province of Almeria; and the third is the most extensive region with Mediterranean climate that occupies the greater part of the Iberian Peninsula. The spatial division basically matches the 800-mm isohyet separating the humid zone from the Mediterranean ones and 300-mm that delimit the southeast semi-arid area.

The first outstanding consideration about the cartographic representation of this central tendency measures is that the general precipitation patter does not change much for most of the country. Nevertheless, there areas some areas where comparing the robust (median) to the non-robust (mean) measures it can be stated that there is a general overestimation of precipitation. The map of the difference between mean and median shows the predominance of greenish tonalities which represent mean precipitation values above medium precipitation (see Fig. 4c). This is a consequence of the positive symmetry of the annual precipitation distributions due to the presence of very rainy years with respect to the rest. The areas where precipitation is overestimated are mainly located in the south normally characterized by low levels of precipitation.

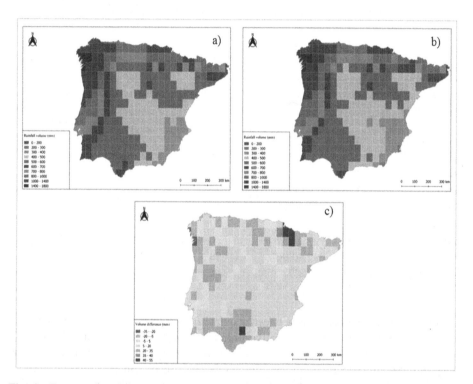

Fig. 4. Cartography of annual central tendency statistics (1901–2016); (a) Mean precipitation, (b) Median precipitation, (c) Difference between mean and median measures [1].

Once the basic robust and non-robust statistics where calculated the number of cell were counted for each precipitation interval. Table 5 shows the results for the central tendency measures of the mean and median, the percentage they represent with respect to the total of cell and the difference between both statistical measures in terms of cells.

Despite of the fact that the differences in the number of cells are not numerous, it strongly indicates that it might be increase significantly when considering finer time scales of analysis.

In relation to the non-robust dispersion statistics (standard deviation) and robust (interquartile range) the most characteristic of both maps is the presence of a marked NW-SE gradient (see Fig. 5) related to the general atmospheric circulation flows.

For these measures the cartographic representation shows significant differences demonstrating the effect of extreme values on the representativeness of dispersion statistical measures. This is one of the basic characteristics of Mediterranean climates so it is relevant for an accurate climatic characterization. In this case, great part of the Iberian Peninsula mainly the center, part of the south and southeast present dispersion values are less than 150 mm linked to lower precipitation records. Therefore, where the annual totals are higher a greater dispersion is observed in the precipitation values; the standard deviation map show higher values than the interquartile range except for the humid zones of the north Atlantic coast characterized by regular rain.

Table 5. Absolute and relative statistical measures of central location by cells.

Precipitation (mm)	Mean cells	Mean cells (%)	Media cells	Media cells (%)	Mean minus Median cells
200–300	0	–	2	0'71	−6
300–400	19	6'71	21	7'42	−2
400–500	64	22'61	70	24'73	−6
500–600	82	28'98	77	27'21	5
600–700	28	9'89	24	8'48	4
700–800	19	6'71	20	7'07	−1
800–1000	35	12'37	36	12'72	−1
1000–1400	31	10'96	28	9'89	3
1400–1800	5	1'77	5	1'77	0

In the map of the differences between standard deviation and interquartile range the latter concentrating the greatest negative differences in the Atlantic coast. These differences indicate the heterogeneity of the precipitation values as a function of their mean and median, which shows the high irregularity in certain areas of the Iberian Peninsula. This is a surprising result since these zones are usually characterized by the regularity of the rainfall regime indicating that there is a significant underestimation of variability in these areas.

Finally, in order to remove the effect con magnitude of precipitation totals, relative dispersion statistics (Pearson Coefficient of Variation and Coefficient of Interquartile Variation) were calculated to be able to compare different zones of the peninsula. In the first map (see Fig. 6) there can be observed a considerable decrease from north to south and even in most of the Mediterranean coast.

The map presents considerable geographic coherence and identifies the area with the most rainfall contrast. Also the cartographic representation changes quite significantly from robust to non-robust statistical measures. The country becomes distinctly divided in two main sectors. First, the northern part with temperate climate due to the influence of Atlantic cyclones and the orographic effect that produce more regular rainfalls and the rest of the territory. Secondly, the Mediterranean extensive sector with highly contrasting precipitation amounts sometimes very large in contrast to others with scarce annual totals.

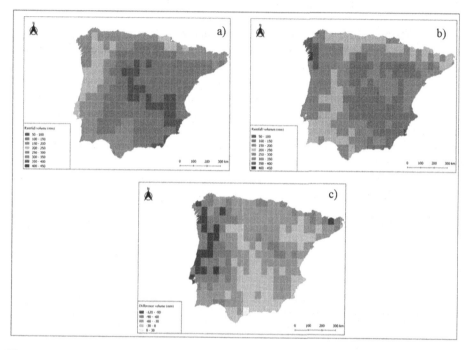

Fig. 5. Cartography of annual absolute statistical dispersion (1901–2016); (a) Annual standard deviation, (b) Annual interquartile range, (c) Difference between the standard deviation and interquartile range measures [1].

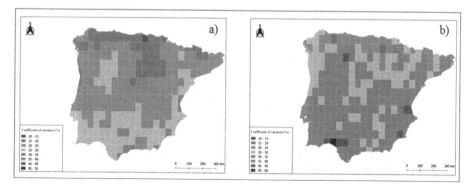

Fig. 6. Cartography of annual relative statistical dispersion (1901–2016); (a) Annual coefficient of variation, (b) Annual interquartile coefficient of variation.

Though the Pearson Coefficient of Variation (CV) has been already used for pre-cipitation studies in the Iberian Peninsula [21], the Coefficient of Interquartile Variation has not been used previously (see Fig. 6b). It shows a less defined pattern than the Pearson's coefficient of variation presenting much higher values of variability. The most notable is the low variability, with respect to the rest of the territory registered in the northern part as well as the presence of higher levels in the Mediterranean coast and the Southwest area of Atlantic influence decreasing as it penetrates the interior of the peninsula.

In the Mediterranean climates precipitation annual totals hide significant variations since the intra-annual variability regime is characterized by strong differences between seasons. While scarcity defines summer season, precipitation is mainly concentrated in winter and spring. However, this general pattern is also highly variable at different spatial scales. Since only mean values are the only statistical measures used in hydrological models to characterize the rainfall regime, it is important to evaluate at this time scale the differences between robust and non-robust statistical measurements. For this reason we will present the seasonal results of the estimated central trend measures according to the previously described procedure.

The following figure shows the maps for the mean, median and difference between these two statistics for winter rainfall (see Fig. 7).

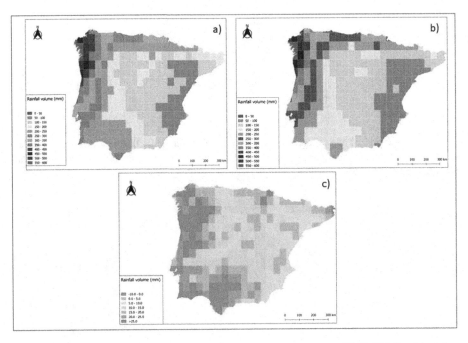

Fig. 7. Cartography of winter central tendency measurements (1901–2016); (a) Mean (b) Median (c) Difference between the mean and the median.

In general, the spatial pattern of spring precipitation clearly reflects the greater rainfall amounts of the Atlantic facade influenced by the extratropical cyclone associated with polar front. In the rest of the territory, characterized by a Mediterranean climate, the higher rainfall is concentrated in the mountain systems of the Iberian Peninsula. In this case, differences between the robust and non-robust measurements are remarkable, especially in the West and Southwest, showing a general overestimation of the precipitation greater than 25 mm (around 12%).

For spring season, the clear northwest-southeast pattern that leaves the rains concentrated on the north Atlantic facade and the Pyrenees mountainous system is evident in the mean (see Fig. 8a) and median maps (see Fig. 8b). In this case the differences between the statistics values are less marked and the difference map shows a slight overestimation of precipitation. Nevertheless, these positive values are more pronounced in the center and in the eastern façade characterized by the highest rates of irregular distribution of precipitation (see Fig. 6b) and many water resources conflicts. Spring precipitation amount here is modest, around 100 mm, so an overestimation of 20 mm represents 20–25% of the water incomes for spring. Even more, spring season is the second pluviometric maximum of the year and represents the latest water contributions before the dry and hot summer, so these amounts are vital for the environmental system.

For spring season, the clear northwest-southeast pattern that leaves the rains concentrated on the north Atlantic facade and the Pyrenees mountainous system is evident in the mean (see Fig. 8a) and median maps (see Fig. 8b).

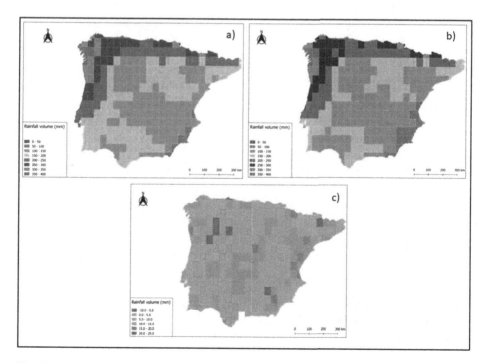

Fig. 8. Cartography of spring central tendency measurements (1901–2016); (a) Mean (b) Median (c) Difference between the mean and the median.

In this case the differences between the statistics values are less marked and the difference map shows a slight overestimation of precipitation. Nevertheless, these positive values are more pronounced in the center and in the eastern façade characterized by the highest rates of irregular distribution of precipitation (see Fig. 6b) and many water resources conflicts.

Spring precipitation amount here is modest, around 100 mm, so an overestimation of 20 mm represents 20–25% of the water incomes for spring. Even more, spring season is the second pluviometric maximum of the year and represents the latest water contributions before the dry and hot summer, so these amounts are vital for the environmental system.

Regarding summer season, the maps show the clear north-south pattern that leaves the rains concentrated on the north Atlantic facade and the Pyrenees mountain range (see Fig. 9a, b). Again the differences are not very noticeable although there is a moderate estimate of resources in the Catalonia region (around 10%). However, summer rainfall in Mediterranean media is always difficult to analyze given its predominantly convective nature that causes a high spatial variability. Therefore, the spatial resolution of this study is not the most appropriate scale of analysis.

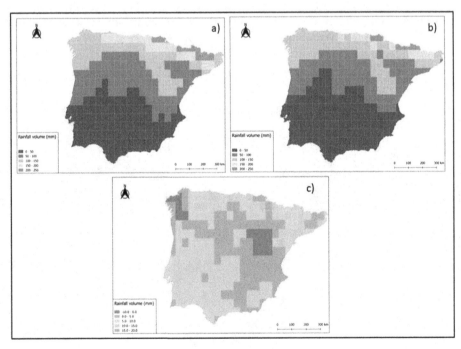

Fig. 9. Cartography of summer central tendency measurements (1901–2016); (a) Mean (b) Median (c) Difference between the mean and the median.

Finally, for the autumn season the spatial differences between the two statisticians are more pronounced and respond to a clear pattern of overestimation in the western façade, the southwest and the Ebro river valley (see Fig. 10). It is the only season in which a slight underestimation is register mainly in the north and southeast part of the territory. Rainfall in autumn is also difficult to estimate at these scales, especially on the Mediterranean façade because they are caused by torrential episodes of highly concentrated rain.

In general, the differences observed in these maps are not depreciable since in many areas, such as the southwest of the Peninsula that corresponds to the Guadalquivir valley, 80% of the water resources are destined to the demand of irrigated agriculture. The expected changes due to climate change, with increased temperatures, evapo-transpiration and extreme events threaten a management system that has already exceeded the natural limits of the resource. This result is particularly relevant as it is there where major imbalances and drought problems occur in the areas where a precise estimation of water resources is most needed in order to be prepared for climate change adaptation and mitigation measurements.

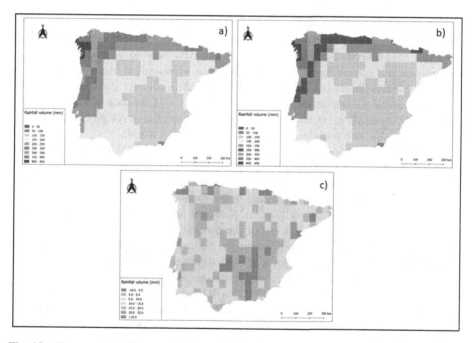

Fig. 10. Cartography of autumn central tendency measurements (1901–2016); (a) Mean (b) Median (c) Difference between the mean and the median.

In addition, the simpler and more efficient way to show all the results obtained in this work is to make them accessible in a geoviewer. This was made by using QGIS Cloud, a web geoviewer developed as an extension of the QGIS that allows the publication in a network of maps, data and geographic services. This geoviewer has all the potential of a cloud storage and broadcast system that provides such a spatial data infrastructure.

In short, it is a platform through which all information and data capable of owning a geographic component can be shared, according to the standards of the Open Geospatial Consortium (OGC), represented on the web through Web Map Services and downloadable in Web Feature Service format. Results of this study can be viewed in https://qgiscloud.com/Juan_Antonio_Geo/expo.

5   Conclusions

The complexity of the spatial rainfall pattern in the Iberian Peninsula is determined by many factors was such as the relief, the layout, orientation or altitude atmospheric circulation. Apart from traditional analogical maps it is difficult to find a cartographic representation based on updated long climatic data and with spatial continuity. Thanks to the availability of global databases is becoming possible. Nevertheless, the complex formats of most of these databases (NetCDF) make still difficult the use of them for non-expert users.

Using the data from the Global Climate Monitor it is possible to carry out this type of studies. This climatic data geo-visualization web tool can greatly contribute to provide an end-user tool for climatic spatial patterns discovery. Compared to other geo-viewers it hast objective advantages such as the easy way to access and visualize climatic past and present (near real-time) data, fast visualization response time, variables and climatic indicators and recently incorporated climatic graphs such as tendency representations and indicators. The fast and easy way for data downloading and exportation in some different accessible formats facilitates the development of climatic studies such as the one presented here for any part of the world.

Using the precipitation data series provided by the GCM, robust and non-robust statistics were calculated for the period 1901–2016 at an annual and seasonal scale. Robust statistical measures provided different and complementary knowledge of precipitation spatial distribution and patterns in the Iberian Peninsula revealing a general significant overestimation of precipitation. Seasonal analysis shows that the overestimation is greater in winter when most of the precipitation amounts fall in the annual regime. The percentages of overestimation are not negligible with values between 10 and 20%. The most worrisome impacts of these differences might occur in spring when the last major rains before the harsh Mediterranean summer occur.

The next basin hydrological plans (2021) will have to incorporate climate change in the future estimation of water resources but the modelled estimation of water resources is still based on the non-updated reference series (1940–2006) as indicated by the Spanish Hydrological Planning Instruction (https://www.miteco.gob.es/es/agua/legislacion/Marco_normativo_planificacion.asp). The results of the present study based on the 1901–2016 period demonstrates the need of reconsidering this reference period as well as the statistical measurements used form these estimations.

The consequences of this for water resources estimation and allocation are noteworthy for environmental management and water planning and should be taken into account. The two coefficients of variation used emphasized that characteristically high irregularity of precipitation in the Iberian Peninsula is higher than it has been traditionally considered and reveals different zones of maximum variability. It is remarkable the differences in the Guadalquivir river valley along the Mediterranean cost where water scarcity and over water demand are major problems causing often water conflicts. The over estimation of water availability together with underestimation of variability could increase this water supply and management problems. In the context of climate change it is even more important to consider this evidences as it might reduce the adaptation capacity.

The most novel contribution of this work is to incorporate new statistical measures and the comparison between them. The results show important variations in the estimation of the amount of the rainfall incomes for water resources. The estimation of the differences between statistics at monthly scales and by river basins is to be achieved. All these results can be useful in the knowledge of spatial and temporal precipitation distribution and, therefore, in the initial computations of the available water resources.

In addition, web-based GIS and geovisualization enable that the results obtained can be displayed in maps and seen in a new web geoviewer (https://qgiscloud.com/ Juan_Antonio_Geo/expo). This is largely novel since usually obtained results in this type of studies are not shared by means of any tool to give greater diffusion through the web. Improving the interface and making it more user-friendly based on the experience of users is a constant goal of the Global Climate monitor project.

In future we would like to present and evaluate the calculated statistics for the rest of the seasonal statistical measurements and monthly scale results for the Iberian Peninsula. Also, this study can be extended to characterization of large areas of the planet being of great interest to compare the evolution and changes of precipitation amounts between different standard periods (climatic normal) and other climatic indicators. So far other climatic variables have been incorporated into the Global Climate Monitor geoviewer and other graphic displays such as climatic diagrams, tendency graphs that can be useful for climatologists, hydrologists, planners and non-experts users such as media workers, policymakers, non-profit organizations, teachers or students. This is an added value to this work concerning not only about the generation of quality climate information and knowledge, but also making it available to a large audience.

References

1. Alfonso Gutiérrez, J., Aguilar-Alba, M., Camarillo-Naranjo, J.M.: GIS and geovisualization technologies applied to rainfall spatial patterns over the iberian peninsula using the global climate monitor web viewer. In: Proceedings of the 4th International Conference on Geographical Information Systems Theory, Applications and Management, GISTAM, vol. 1, pp. 79–87 (2018). https://doi.org/10.5220/0006703200790087. ISBN 978-989-758-294-3
2. Cabello Villarejo, V., Willaarts, B.A., Aguilar-Alba, M., Del Moral Ituarte, L.: River basins as socio-ecological systems: linking levels of societal and ecosystem metabolism in a mediterranean watershed. Ecol. Soc. **20**(3), 20 (2015)
3. Camarillo-Naranjo, J.M., Limones-Rodríguez, N., Álvarez-Francoso, J.I., Pita-López, M.F., Aguilar-Alba, M.: The Globalclimatemonitor system: from climate data handling to knowledge dissemination. Int. J. Digit. Earth (2018). https://doi.org/10.1080/17538947. 2018.1429502

4. Carslaw, D.C., Ropkins, K.: Openair—an R package for air quality data analysis. Environ. Model Softw. **27–28**, 52–61 (2012)
5. Casado, M.J., Pastor, M.A., Doblas-Reyes, F.J.: Links between circulation types and precipitation over Spain. Phys. Chem. Earth, Parts A/B/C **35**(9), 437–447 (2010)
6. Climate Research Unit (n.d.). British Atmospheric Data Center. http://www.cru.uea.ac.uk/data
7. Cortesi, N., González-Hidalgo, J.C., Trigo, R.M., Ramos, A.M.: Weather types and spatial variability of precipitation in the Iberian Peninsula. Int. J. Climatol. **34**(8), 2661–2677 (2014)
8. De Castro, M., Martin-Vide, J., Alonso, S.: El clima de España: pasado, presente y escenarios de clima para el siglo XXI. Impactos del cambio climático en España. Ministerio de Medio Ambiente, Madrid (2005)
9. Edwards, P.N., Mayernik, M.S., Batcheller, A.L., Bowker, G.C., Borgman, C.L.: Science fiction: data, metadata, and collaboration. Soc. Stud. Sci. **41**, 667–690 (2011). https://doi.org/10.1177/0306312711413314
10. Folland, C.K., et al.: Observed climate variability and change. In: Climate Change 2001: The Scientific Basis. Contribution of Working Group I to the Third Assessment Report of the Intergovernmental Panel on Climate Change, pp. 99–181. Cambridge University Press, Cambridge (2001)
11. García-Barrón, L., Aguilar-Alba, M., Sousa, A.: Evolution of annual rainfall irregularity in the southwest of the Iberian Peninsula. Theoret. Appl. Climatol. **103**(1–2), 13–26 (2011)
12. García-Barrón, L., Morales, J., Sousa, A.: Characterisation of the intra-annual rainfall and its evolution (1837–2010) in the southwest of the Iberian Peninsula. Theoret. Appl. Climatol. **114**(3–4), 445–457 (2013)
13. García-Barrón, L., Aguilar-Alba, M., Morales, J., Sousa, A.: Intra-annual rainfall variability in the Spanish hydrographic basins. Int. J. Climatol. (2017). https://doi.org/10.1002/joc.5328
14. Global Precipitation Climatology Centre (n.d.) Product Access. http://www.dwd.de/
15. Hidalgo-Muñoz, J.M., et al.: Trends of extreme precipitation and associated synoptic patterns over the southern Iberian Peninsula. J. Hydrol. **409**(1), 497–511 (2011)
16. Intergovernmental Panel on Climate Change. Climate Change 2013 – The Physical Science Basis Working Group I Contribution to the Fifth Assessment Report of the Intergovernmental Panel on Climate Change (2014). http://www.ipcc.ch/report/ar5/wg1/
17. Jones, P.D., Moberg, A.: Hemispheric and large-scale surface air temperature variations: an extensive revision and an update to 2001. J. Clim. **16**, 206–223 (2003)
18. Jones, W.R., Spence, M.J., Bowman, A.W., Evers, L., Molinari, D.A.: A software tool for the spatiotemporal analysis and reporting of groundwater monitoring data. Environ. Model Softw. **55**, 242–249 (2014)
19. Krysanova, V., et al.: Cross-comparison of climate change adaptation strategies across large river basins in Europe, Africa and Asia. Water Resour. Manag. **24**(14), 4121–4160 (2010)
20. López-Bustins, J.A., Sánchez Lorenzo, A., Azorín Molina, C., Ordóñez López, A.: Tendencias de la precipitación invernal en la fachada oriental de la Península Ibérica. Cambio Climático Regional y Sus Impactos, Asociación Española de Climatología, Serie A, (6), 161–171 (2008)
21. Martín-Vide, J.: Estructura temporal fina y patrones espaciales de la precipitación en la España peninsular. Memorias de la Real Academia de Ciencias y Artes de Barcelona, 1030, LXV, vol. 3, pp. 119–162 (2011)
22. Martin-Vide, J.: Patrones espaciales de precipitación en España: Problemas conceptuales. In Clima, ciudad y ecosistema. In: Fernández-García, F., Galán, E., Cañada, R. (eds.) Asociación Española de Climatología Serie B, no. 5, pp. 11–32 (2011)

23. Mitchell, T.D., Jones, P.D.: An improved method of constructing a database of monthly climate observations and associated high-resolution grids. Int. J. Climatol. **25**, 693–712 (2005)
24. Muñoz-Díaz, D., Rodrigo, F.S.: Seasonal rainfall variations in Spain (1912–2000) and their links to atmospheric circulation. Atmos. Res. **81**(1), 94–110 (2006)
25. New, M., Hulme, M., Jones, P.D.: Representing twentieth century space–time climate variability. Part 2: development of 1901–96 monthly grids of terrestrial surface climate. J. Clim. **13**, 2217–2238 (2000)
26. Ríos-Cornejo, D., et al.: Links between teleconnection patterns and precipitation in Spain. Atmos. Res. **156**, 14–28 (2015)
27. Vitolo, C., Elkhatib, Y., Reusser, D., Macleod, C.J.A., Buytaert, W.: Web technologies for environmental Big Data. Environ. Model Softw. **63**, 185–198 (2015)

What Are You Willing to Sacrifice to Protect Your Privacy When Using a Location-Based Service?

Arielle Moro[✉] and Benoît Garbinato

Department of Information Systems, University of Lausanne, Lausanne, Switzerland
{arielle.moro,benoit.garbinato}@unil.ch

Abstract. Today, we use location-based services on a daily basis. They provide information related to the current location of the users and are extremely helpful. The next step of location-based services is to use predicted locations of users to create new content or to improve the quality of existing ones. Today location-based services must capture a large location history of the users in order to be able to build predictive mobility models of users and forecast their future locations. This is a clear privacy issue because these services are thus able to also obtain sensitive information related to users. In this paper, we propose a system that provides future locations of users to location-based services, protects the location privacy of the users with spatio-temporal conditions and ensures the utility of the information provided by the location-based services. The user is a key actor of the system and is involved in the protection process because she indicates these spatio-temporal conditions, which express the spatio-temporal utility she is willing to sacrifice in order to protect her privacy. We evaluate the two components of the system according to two perspectives: a prediction accuracy analysis and a utility/location privacy evaluation. The proposed system provides satisfactory prediction accuracy results that exceed 70%. The utility/privacy evaluation shows that our mechanism obtains the best results in terms of utility and location privacy compared to two other common location privacy preserving mechanisms. Hence, these evaluations confirm the relevance of our system.

1 Introduction

Nowadays, location-based services are used on a daily basis. They became popular due to the democratization of the mobile devices. Location-based services help users by displaying location-based information on their mobile devices by using locations captured by the positioning system included in the devices. For example, a location-based service can provide the next departures of public transportation located in the vicinity of the user's current location. These services only take into account the current location of the user. The next step of location-based services will include the prediction of locations of the user in order to create new content or improve their existing content to display on the mobile device of the

© Springer Nature Switzerland AG 2019
L. Ragia et al. (Eds.): GISTAM 2018, CCIS 1061, pp. 108–132, 2019.
https://doi.org/10.1007/978-3-030-29948-4_6

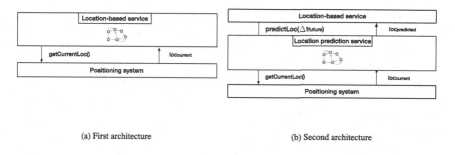

(a) First architecture   (b) Second architecture

Fig. 1. Problem statement overview through two different system architectures.

user. For instance, a location-based service will be able to display personalized information on the user's mobile device in advance, such as the menus of different restaurants in the vicinity of a location in which she will probably be at a specific time, e.g., Monday between 11:30 am and 12:00 pm. After the prediction, the location-based service will display these menus in advance, e.g., Monday at 11:00 am.

In order to predict future locations of users, the research community showed that mobility prediction requires a large amount of locations to create reliable predictive models. Today, existing mobile platforms (Android, iOS) do not provide location predictions of users via their Application Programming Interfaces, i.e., APIs. They only provide the access to the current location of a user. This means that location-based services frequently obtain locations of the user from the positioning system, located on the trusted part of the mobile device, in order to gather a rich location history and to create her predictive model as depicted in Fig. 1(a). This predictive model can be stored on the mobile device of the user or in a cloud service that can be seen as an extension of the location-based service. Consequently, this is clearly a privacy issue because the location-based service can be an adversary and has access to the entire location history of the user. This adversary could try to infer personal locations of the user (e.g., work place, home place, favorite places) and personal information derived from her locations (e.g., her friends, her religion, her political choices). In [3], Backes et al. present a novel social inference attacks that aims at summarizing mobility features of users and, more specifically, the social links amongst them. In the literature, Krumm shows that sharing accurate locations has a real cost for a user because a potential adversary cannot only discover a lot of sensitive information related to the user but also identify her by just performing simple localization attacks as described in [13]. Moreover, Zang and Bolot highlight that a few number of user's locations only might highly compromise the location privacy of a user in [18]. In order to mitigate this privacy issue, one could propose to simply enrich the operating system layer of the mobile devices by adding a predictive service, which is presented in Fig. 1(b). In this architecture, the location-based service obtains future locations of users from the trusted part of the mobile device. This architecture would enable to protect the location history of the user and her predictive model. However, it would be possible for the location-based service to partially or entirely rebuild the predictive model of the user after a large number of requests. Consequently, this solution would partially preserve the user location privacy.

In this paper, we present a system that enables to provide future locations of a user to a location-based service while preserving the location-privacy of the user and ensuring a good utility of the service. This system contains two components: a location prediction component, which contains a new predictive model called a *location trend model* and a location privacy component, which uses a new Location Privacy Preserving Mechanism (LPPM) called a *utility/privacy tradeoff*. The user is a key actor of our system because she must indicate the temporal and spatial sacrifices she is willing to do in order to protect her location privacy. It is important to mention that this paper is an extended version of a paper previously accepted and presented in a conference [14]. We evaluate the location prediction component of our system from a prediction accuracy perspective in a binary manner and in a distance accuracy manner. Then, we also evaluate the location privacy component from a utility/location privacy (uncertainty and accuracy) perspective by comparing it to two other LPPMs of the literature, i.e., the Gaussian perturbation and the grid-based rounding. Regarding the location privacy evaluation, we compute the level of uncertainty of an adversary that performs a localization attack on some of predicted locations received from our system. The metric used to compute this uncertainty is based on the well-known Shannon entropy. For these evaluations, we use a dataset called *Breadcrumbs*, which was the result of a data collection campaign of three-months that we conducted (from the end of March 2018 to the end of July 2018). From this dataset, we selected rich location datasets of 47 users. The results obtained for the evaluation of these two components are satisfactory and confirm the relevance of the system. The prediction accuracy of the model can exceed 70% and we obtain the best utility/privacy results for our LPPM. The contributions of this paper are listed below.

- We describe a system that allows a location-based service to request future locations of a user;
- We present a new statistical predictive location model that contains location trends of a user organized by time slices;
- We describe a new location privacy preserving mechanism that enables to reach an appropriate utility/privacy tradeoff.

The paper has the following structure: we first recall the problem statement of the paper in Sect. 2. Section 3 presents the system model and the adversary model of this paper. The proposed system is entirely described in Sect. 4. Then, we detail the evaluations of the system from a prediction accuracy point of view and a utility/location privacy perspective in Sect. 5. We describe the closest works of the literature to the two main subjects of this paper in Sect. 6, which are the location prediction as well as the location privacy. Finally, we recall the most important findings of the paper and present the future work in Sect. 7.

2   Problem Statement

If we consider that a location-based service needs future locations of a user to provide a location-based content to her in advance, the service must create a predictive model of the user. In order to reach this goal, the location-based service must frequently obtain locations of the user in order to update her predictive model as shown in Fig. 1(a). This is a clear location privacy issue because all the locations of the user are regularly shared with the location-based service. Thus, this means that a large number of sensitive information linked to the user is given to a possible adversary. For example, the location-based service can discover personal locations and sensitive derived information from the locations of the user, such as her home and work places but also her religious and political choices.

In order to reduce the location privacy issue, we could create a dedicated service at the operating system layer that we consider as trusted, as shown in Fig. 1(b). In this system architecture located on the mobile device of the user, the only service that has access to the raw locations of the user coming from the positioning system is the dedicated service. This service provides predicted locations to the location-based service that needs them to operate properly. Although we mitigate the location privacy risk in this context, there is still another location privacy issue regarding the predicted locations shared with the location-based service. According to the number of requests performed by the location-based service, this latter will be able to rebuild a partial or maybe an entire view of the predictive model of the user.

The challenge is thus to protect as much as possible the location privacy of the user while ensuring a good quality of service, i.e., the utility of the location-based service perceived by the user. Although there exist various LPPMs in the literature, they do not necessarily meet the utility requirement of a location-based service and do not include the user as a key actor of the system. Indeed, already proposed LPPMs can easily compromise the use of the location-based service because they are not automatically adapted to the temporal and spatial utility constraints of the service. The location-based service can thus quickly become unusable. For example, the location information provided by the location-based service can be inaccurate or simply erroneous because the precision of the predicted location has been made too low by the LPPM. As a result, the user might simply stop using the location-based service. As discussed in the introduction, our approach consists in building a system that takes into account the utility requirement of a location-based service and temporal and spatial conditions expressed by the user in order to protect her location privacy.

3   System and Adversary Models

This section describes the key definitions that are crucial to present before the description of the system. First, we introduce the definitions of a user and her location history. Second, we present the adversary model of the system, which is important in a location privacy context.

User and Location History. A user is a human who moves on a two-dimensional space and owns a mobile device. This device is able to record the locations of the user and when they are detected via a positioning system that uses GPS, WiFi or radio cells to locate the user. A location is described as a triplet $loc = (\phi, \lambda, t)$, in which ϕ and λ are the latitude and longitude of the location in the two-dimensional space and t is the time when the location was captured by the positioning system. The location history of a user, contained in the mobile device, is regularly updated by the mobile device with the new locations coming from the positioning system. The location history of the user is a sequence of locations $L = \langle loc_1, loc_2, \cdots, loc_n \rangle$. We can express the latitude, longitude and time of a location loc_i by directly writing $loc_i.\phi$, $loc_i.\lambda$ and $loc_i.t$ respectively.

Adversary Model. The adversary model of this paper takes into account an honest but curious adversary that is the location-based service. This adversary uses our system to obtain future locations of the user. After multiple requests for a specific Δt in the future, the adversary will try to infer one single location for this future time according to all the locations obtained from our system. The locations received from our system is the unique knowledge of the adversary on which the localization attack is performed. The honest but curious behavior of the adversary means that it will not try to break the sharing protocol or obtain the location predictive model of our system. Moreover, we consider that the location-based service, i.e., the adversary, gives honest values to our system when it requests a new predicted location.

4   System Overview

The system, presented in this paper, contains two component: the first component focuses on the prediction of the future locations of the user and the second component aims at spatially protecting the locations provided by the first component. The structure of the system is presented in Fig. 2. In order to better understand the parameters that have an influence on the process of the system, we will first present them below. Then, we will describe the two components of the system.

Location-Based Service Parameters. A location-based service will request our system in order to obtain the location of a user in Δt_{future} seconds in the future. In addition, the location-based service indicates its required utility $\Delta r_{utility}$ in meters that allows it to operate properly and give a reliable location-based information to the user. For instance, if a location-based service must call a taxi for a user in advance, it will indicate a utility of a short distance in meters, such as 500 m. A long distance could compromise the use of the taxi service itself because it could display inaccurate information to the user. The returned value by our system is a location expressed by a

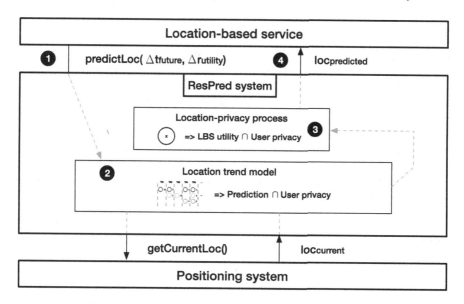

Fig. 2. System overview.

pair $loc_{predicted} = (\phi, \lambda)$. Equation 1 below summarizes the request.

$$loc_{predicted} = predictLoc(\Delta t_{future}, \Delta r_{utility}) \tag{1}$$

To summarize, the system will answer the following question: *Knowing the user will need some location-based information within $\Delta r_{utility}$ meters in Δt_{future} seconds from now, where will be the user?*

User Parameters. The user is a key actor of our system because she can specify two parameters that indicate the temporal and spatial maximum utility percentages she is willing to lose in order to protect her privacy: $\Delta temporal_{maxUtilityLoss}$ and $\Delta spatial_{maxUtilityLoss}$ respectively. These parameters can be set for different location-based services and thus adapted to each service that the user decides to use. The first parameter is used in the location prediction component and the second in the location privacy component. Although the first parameter helps to temporally obfuscate the predicted location, it can only be involved during the prediction process because it has an influence on the future computed time that is used to predict the future location of the user.

4.1   Location Prediction Component

The location prediction component contains a predictive model that is called a *location trend model*. This component aims at predicting a temporary region $tmpRegion_{predicted}$ visited by the user that will be sent to the location privacy component. This temporary visited region is a triplet that contains a latitude, a longitude and a radius, such that $tmpRegion_{predicted} = (\phi, \lambda, \Delta r)$. In order to predict it, the component uses three crucial variables: the current location of the user $loc_{current}$ obtained from the positioning system, the number of seconds in the future Δt_{future} indicated by the location-based service and the percentage $\Delta temporal_{maxUtilityLoss}$ that represents the temporal max utility loss set by the user. The Eq. 2 below summarizes the request handled by the location prediction component.

$$tmpRegion_{predicted} = predictRegion(loc_{current}, \Delta t_{future}, \Delta temporal_{maxUtilityLoss})$$
$$(2)$$

Before presenting the structure of the model, we must define the notions of time slices and visited regions as detailed in the next two paragraphs.

Time Slices. In order to create the *location trend model*, we discretize the time into n time slices over one period, e.g., every 20 min over one week. A time slice is defined as a triplet, such that $ts = (t_s, t_e, index)$, in which t_s (e.g., Monday - 7:00 am) and t_e (e.g., Monday - 7:20 am) represent the starting time and ending time of the time slice and *index* is its unique identifier ranging between 1 and n, both included. For instance, if we generate time slices that have a duration of 20 min during a period of 1 week, we will obtain 504 time slices. All the computed time slices over one period can be represented as a sequence called *timeslices*, such that $timeslices = \langle ts_1, ts_2, \cdots, ts_n \rangle$.

Visited Regions. A visited region is a circular region that was visited by the user during one time slice. More formally, a visited region vr is defined as a quadruplet, such that $vr = (\phi, \lambda, \Delta r, c)$, in which ϕ and λ describe the center of the visited region, Δr is its radius expressed in meters and c is a counter of visits. When a visited region is created, the counter is automatically set to 1. In order to compute a visited region, we consider a subsequence of the location history L of the user that contains n successive locations l_{sub} that occurred during a specific time slice, such that $l_{sub} = \langle loc_1, loc_2, \cdots, loc_n \rangle$. The center of this visited region is the barycenter of all the locations of the subsequence and the radius corresponds to the distance between the center and the farthest location of the subsequence.

Structure of the Location Trend Model

The *location trend model* has a structure that follows the number of time slices created over a period. The structure of the model is circular because we consider the last time slice of the model is connected to the first time slice and conversely. Each time slice contains a list of visited regions by the user and a counter of visits associated to each visited region of the time slice as described in Fig. 3.

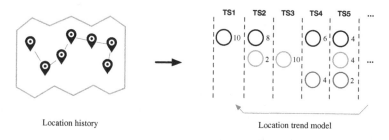

Location history   Location trend model

Fig. 3. Creation of the location trend model.

Update of the Location Trend Model

We first assume that the *location trend model* is empty and, consequently, only contains empty time slices. Then, we start collecting locations of the user and the visited regions of the previous time slice are created or updated when the beginning timestamp of the current time slice is reached. In the case when there are existing visited regions in a time slice when a new one is created, this new visited region is compared with the existing visited regions already computed. The comparison works as follows: if there is an intersection between the new visited region and an existing visited region, they are merged together and a new visited region is created. The counter of visits of this new visited region is the addition of the counters of the two visited region merged. Hence, the model can be updated online in real time with a buffer of locations that depends on the duration of the time slice.

It is important to highlight that we improved the structure of this model compared to the one presented in the previous paper [14]. More specifically, we decided to avoid using sequences of regions of interest visited by a user to fill the time slices of the predictive model because we could miss some crucial and frequent transitions amongst these regions. Consequently, we decided to only focus on the extraction of visited regions during the pre-computed time slices in order to create and update it.

Prediction of a Temporary Predicted Region

In order to predict a temporary region $tmpRegion_{predicted}$, the component first starts computing a current time slice, which can be found with the current location of the user captured by the component. Second, a target time slice is computed according to the current time slice and the Δt seconds in

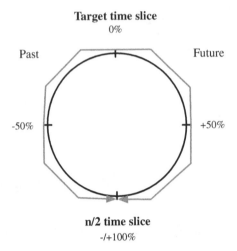

Fig. 4. Possible temporal shifts of a target time slice.

the future indicated by the location-based service. Then, we use the parameter $\Delta temporal_{maxUtilityLoss}$ set by the user to shift the target time slice and bring temporal noise to the predicted value if needed. We assume that this $\Delta temporal_{maxUtilityLoss}$ is computed according to the influence of the time on the location of the user, which is depicted in Fig. 4. Thus, we choose a direction (i.e., past or future) and a random value between 0% and 100% (both included due to the circular characteristic of the temporal shift process). These two elements add temporal noise to the target time slice and, consequently, change the value of the target time slice. As soon as the target time slice is properly defined, we start analyzing its visited regions. We select the visited region that is the most visited by the user, which has the highest visit counter. If two (or more) visited regions have the same visit counter, we select the visited region that was the most recently visited. At the end of the process, we create the $tmpRegion_{predicted}$, which has a center and a radius equal to the center and the radius of the visited region selected previously.

4.2   Location Privacy Component

The goal of the location privacy component is to blur the $tmpRegion_{predicted}$ spatially obtained from the location prediction component and to create new coordinates that will be sent to the location-based service, i.e., $loc_{predicted} = (\phi, \lambda)$. The LPPM used in the system and described hereafter is called a *utility/privacy tradeoff*. Intuitively, this transformation aims at reaching a tradeoff between two crucial parameters that help at protecting as much as possible the location privacy of the user while maintaining a good quality of the location-based service: $\Delta spatial_{maxUtilityLoss}$ percentage that is set by the user and $\Delta r_{utility}$ that is a distance in meters indicated by the location-based service. This component answers the following Eq. 3 expressed below:

$$loc_{predicted} = protect(tmpRegion_{predicted}, \Delta spatial_{maxUtilityLoss}, \Delta r_{utility}) \quad (3)$$

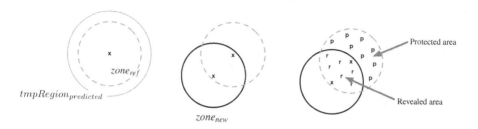

Fig. 5. Location privacy process.

After having received the $tmpRegion_{predicted}$ from the location prediction component, the location privacy component creates a reference zone $zone_{ref}$ that has a center equals to the center of $tmpRegion_{predicted}$ and a radius that corresponds to $\Delta r_{utility}$. It is obvious that the $zone_{ref}$ can become lower or greater than the $tmpRegion_{predicted}$. For this, the choice of the radius of the visited regions is crucial and closely depends on the chosen duration of the time slices of the *location trend model* of the user. We also consider a new zone $zone_{new}$ that has a center equal to the new coordinates generated by the component and a radius that is equal to $\Delta r_{utility}$. These new coordinates are computed in the following manner. We first generate an angle randomly and new coordinates in the direction of the angle between 0 and a threshold value corresponding to the case where there cannot have any intersection between $zone_{ref}$ and $zone_{new}$, i.e., $2 \times \Delta r_{utility}$. Then, we compute the percentage of area revealed to the location-based service based on the area of the intersection between $zone_{ref}$ and $zone_{new}$, which is called $p_{revealed}$. This percentage is computed as follows in Eq. 4:

$$p_{revealed} = \frac{area(zone_{ref} \cap zone_{new})}{area(zone_{ref})} \quad (4)$$

With this revealed percentage, we are now able to compute a protected percentage called $p_{protected}$ that describes the percentage of the protected area of the $zone_{ref}$, which is simply equal to $1 - p_{revealed}$ (scaled to 1). Figure 5 depicts the entire process that includes the temporary predicted region $tmpRegion_{predicted}$ (gray circle), the reference zone $zone_{ref}$ (gray circle with a dotted line), the new zone $zone_{new}$ (black circle), the revealed area of the reference zone (zone with the symbols "r") and the protected area of the reference zone (zone with the symbols "p") that enable to compute the revealed and the protected percentages.

Now, we use the parameter given by the user in order to protect her location privacy, which is $\Delta spatial_{maxUtilityLoss}$ in order to evaluate if the protected percentage $p_{protected}$ is lower or equal to $\Delta spatial_{maxUtilityLoss}$, as described in Condition 5 below. The spatial maximum utility loss percentage $\Delta spatial_{maxUtilityLoss}$ can have a value ranged between 0% (included) and 100% (not included because there cannot have any intersection between the two zones). If the condition is met, we create $loc_{predicted}$ that has a latitude and a longitude that correspond to the latitude and the longitude of the center of the new zone $zone_{new}$. Finally, $loc_{predicted}$ is sent to the location-based service.

$$p_{protected} <= \Delta spatial_{maxUtilityLoss} \tag{5}$$

5   Evaluation

This section aims at presenting the evaluations of the two components that belong to the system. First, we detail the dataset used for the evaluation of the system. Second, we present an evaluation of the location prediction component from a prediction accuracy perspective. Third, we show an evaluation of the location privacy component from a utility/location privacy perspective.

5.1   Breadcrumbs Dataset

The *Breadcrumbs* dataset[1] was chosen to evaluate the system. During a period of approximately three months (from the end of March 2018 to the end of June 2018), the locations as well as additional data of 81 participants were collected by our research team *DopLab*. In the following evaluations, we only use the locations of the participants. All participants were students who belonged to the University of Lausanne or Swiss Federal Institute of Technology of Lausanne (EPFL). An iOS application was given to the participants in order to frequently capture their locations on the basis of one location per hour in the best possible case. From this dataset, we only select rich location datasets of 47 participants that have the following characteristic: a number of days without any location lower or equal to 5 days. In addition, it is important to indicate that the percentage of hours per day having at least one location per hour is greater than 43% for all participants and that this percentage is greater or equal to 70% for 26 participants. Instead of using the word *participant*, we use the word *user* further in order to be aligned with our system model described in Sect. 3.

5.2   Evaluation of the Location Prediction Component

The goal of this section is to evaluate the reliability of the location prediction of the *location trend model* from different angles. We first present the evaluation method and then the obtained results about the prediction accuracy of the model. In addition, we describe an evaluation of the temporal utility loss according to different shifts of the target time slice.

---

[1] Breadcrumbs data collection campaign website: https://bread-crumb.github.io.

Methodology of the Location Prediction Evaluation

We split the dataset, i.e., location history, of each user in two distinct parts: the training part and the test part. We assume that the locations of each user dataset are ordered in a chronological order according to their timestamp. The training part is equal to the first 80% of the user dataset and the test part corresponds to the last 20% of the user dataset. We follow the process described in Sect. 4.1 in order to create the *location trend model*.

We decide to compute a *binary* evaluation and an *accuracy in meters* evaluation. In order to evaluate the location prediction accuracy, the test part of locations of a user is divided into multiple subsequences of locations that all have a duration equal to that of the temporal granularity of the computed time slices of the *location trend model*. We predict a $tmpRegion_{predicted}$ for each time slice of the time slice sequence of the test part and check if at least one location of the subsequence of the time slice is contained in the $tmpRegion_{predicted}$ related to it. This evaluation is called a *binary* evaluation. Finally, the *accuracy in meters* evaluation is computed for all predictions between the center of the $tmpRegion_{predicted}$ and the nearest location of the subsequence of the time slice.

Results of the Location Prediction Evaluation

The *binary* evaluation and the *accuracy in meters* evaluation are depicted in Figs. 6 and 7. In these figures, we test several time slice durations, i.e., 144, 72, 48, 24, 12 and 8 slices per day over one week that correspond to every 10 min, 20 min, 30 min, 1 h, 2 h and 3 h over one week respectively. For sake of simplicity, we compute the mean of the *binary* evaluation and the *accuracy in meters* evaluation of all 47 selected users of the *Breadcrumbs* dataset. As you can notice in Figs. 6 and 7, the higher the duration of the time slice, the higher the *binary* evaluation results and the *accuracy in meters* evaluation results. Although it is not indicated in the figure, it is important to mention that 11 users have a binary percentage result greater than 70% (scaled to 100%) for a time slice duration of 30 min (i.e., 48) and 20 users for a time slice duration of 1 h (i.e., 24). This means that it could be interesting to adapt the time slice duration to the mobility of the user and choose a time slice duration per user. Moreover, Fig. 8 depicts the evolution of the entropy mean of the *location trend model* of all users for the different time slice durations. It clearly highlights that the higher the time slice duration, the lower the entropy mean of the *location trend model*. The entropy of one time slice e_{ts} is computed with the well-known Shannon entropy equation that takes into account the probability p_{vr_i} to visit each visited region of one time slice as described in Eq. 6 below (n is the total number of visited regions in one time slice). In order to obtain the final entropy of the *location trend model* of one user, we compute the mean of the entropy results obtained for all time slices over one week. Then, we finally compute the mean of the final entropy of all users in order to create Fig. 8.

$$e_{ts} = -\sum_{i=1}^{n} p_{vr_i} \log_2 p_{vr_i} \qquad (6)$$

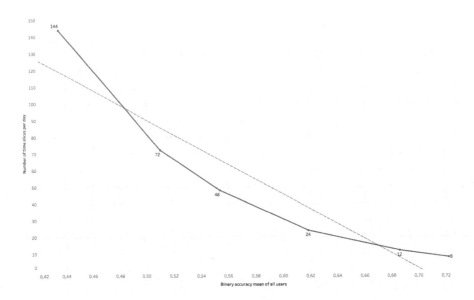

Fig. 6. Evolution of the binary accuracy mean of all users.

In addition, it was interesting to see the evolution of the mean of the radius of the visited regions of all users and the mean of the number of visited regions of all users according to the different time slice durations. In Fig. 9, we see that the mean of the radius of the visited regions of all users increases with the duration of the time slices. However, the mean of the number of visited regions of all users according to the different time slice durations does not follow the increase of the duration of the time slices as it is depicted in Fig. 10. This is probably due to the fact that it depends on the mobility of the users and the duration of the time slices. For this specific *Breadcrumbs* dataset, a very low (i.e., 144) or high (i.e., 8, 12) time slice durations may capture a lower number of visited regions compared to the other durations. This could be linked to the size of the visited regions: very few and small for 144 and very few and big for 8 and 12.

Thus, these results give a good indication of the level of granularity of the time slice we can choose to create the *location trend model*. For this paper, we select the time slice duration in order to reach a tradeoff between a reasonable binary/accuracy in meters prediction and the realistic distance mean in meters of the visited regions. Consequently, we choose a time slice duration of 30 min (i.e., 48) and 1 h (i.e., 24) for the next evaluations even if the second first one is better in terms of prediction accuracy.

Evaluation of the Temporal Utility Loss

As we presented in Sect. 4.1, target time slice can be shifted of one or multiple time slices. This shift is indicated by the user with the percentage $\Delta temporal_{maxUtilityLoss}$. Then, this percentage is converted into a shift according

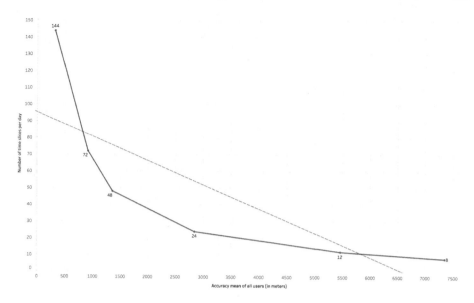

Fig. 7. Evolution of the accuracy mean of all users (in meters).

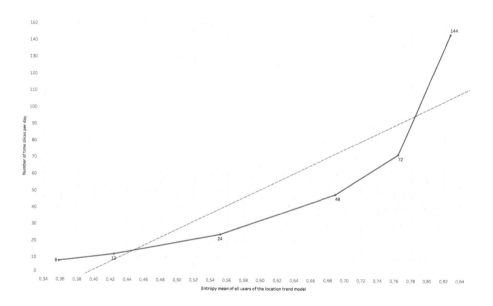

Fig. 8. Evolution of the entropy mean of the location trend model of all users.

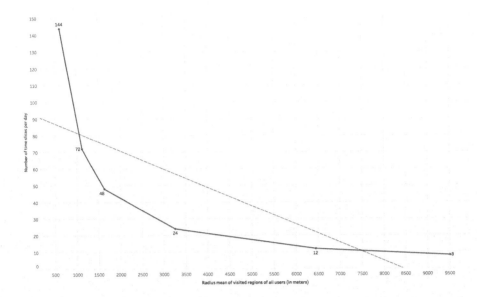

Fig. 9. Evolution of the mean of the radius of the visited regions of all users.

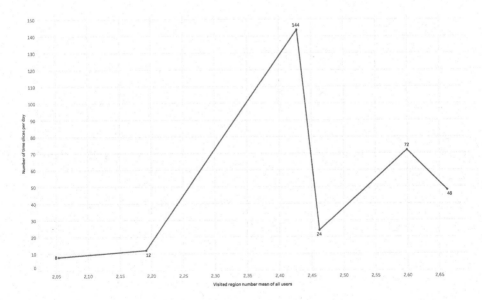

Fig. 10. Evolution of the mean of the number of the visited regions of all users.

to the rules expressed in Fig. 4. However, it is important to show if this assumption is true with real *location trend models* of users. In order to evaluate it, we take a reference visited region for a specific time slice, which corresponds to the most visited region of this time slice. Then, we compute the distance between the center of this reference visited region and the center of the next most visited time slice of each next time slice until reaching again the reference visited region. This evaluation is thus circular and follows an ascendant order. We do this evaluation for each time slice of the *location trend model* of all users. At the end, we simply compute the mean of all the values obtained for each shift (from 1 shift to the total number of possible shifts that is equal to the total number of time slices of the *location trend model*) of all users. Figure 11 presents the results obtained for a model with the time slices that have a duration of 30 min on the left and a duration of 1 h on the right. As we can notice, the rules expressed in Fig. 4 follow the reality depicted in Fig. 11. When we reach the middle of all possible shifts, we obtain the greatest distance compared to the two closest shifts, i.e., the next one and the previous one, that are at the beginning and at the end of each graph.

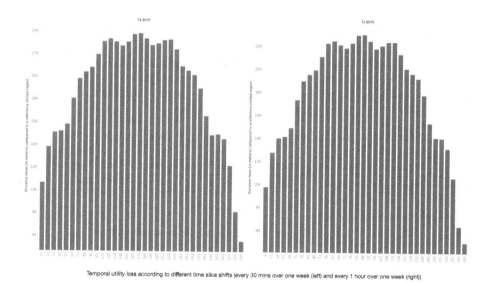

Temporal utility loss according to different time slice shifts (every 30 mins over one week (left) and every 1 hour over one week (right))

Fig. 11. Evolution of the distance in meters according to all time slices.

5.3   Evaluation of the Location Privacy Component

The goal of this section is to evaluate the location privacy component. We compare the *utility/privacy tradeoff* LPPM that is used in the location privacy component of our system to two other LPPMs, which are the Gaussian perturbation and the grid-based rounding. First, we detail the two other LPPMs selected from the literature and their parameters. Second, we describe the scenarios chosen for the evaluation and their parameters. The scenarios are two possible use-cases of our system. Third, the three chosen metrics, which enable to assess the utility and the location privacy, are presented.

Location Privacy Preserving Mechanisms and Parameters

The first chosen LPPM is the Gaussian perturbation described in [2]. This mechanism adds Gaussian noise to the latitude and the longitude of a raw location according to a certain mean and a standard deviation. The selected parameters for this mechanism are four standard deviations that are 0.0005, 0.005, 0.05 and 0.5 (the mean being the latitude or the longitude of the raw location). They are ranged from approximately 55 to 66500 m respectively. The second chosen LPPM is the grid-based rounding presented in [1,13]. This mechanism works with a grid that discretizes space in which the user moves. It modifies the coordinates of a raw location into new coordinates that correspond to the nearest vertex of the cell of the grid in which the raw location is. The parameters related to this LPPM are 0.005, 0.05 and 0.5 between two successive latitudes or longitudes to create each cell of the grid. These values are ranged from approximately 380 to 38000 m. The parameters chosen for the *utility/privacy tradeoff* mechanism that we present in Sect. 4.2 are 0.2, 0.4, 0.6 and 0.8 for the value of $\Delta spatial_{maxUtilityLoss}$. It is important to note that we do not take into account the temporal modification expressed with $\Delta temporal_{maxUtilityLoss}$ in this evaluation, which is used in the location prediction component. Thus, we consider that $\Delta temporal_{maxUtilityLoss}$ is equal to 0%, i.e., there is no temporal shift.

Table 1. List of parameters used for each location-based service scenario.

Parameter/Location-based service scenario	Public transportation scenario	Taxi scenario
Location-based service utility	1000 m	500 m
Number of target time slices	10	4
Number of prediction requests	100	100
Time slice distribution of prediction requests	Randomly or evenly distributed	

Scenarios and Parameters

In order to properly assess our system and, more specifically, the *utility/privacy tradeoff* against the two other LPPMs, we select two different location-based service scenarios. The first scenario is a location-based service that displays public transportation information to a user. For example, this location-based service will display the next departures of bus, train, metro in advance in the vicinity of the location of the user in the future. The second scenario is a location-based service that calls a taxi in advance for a user in the vicinity of the future location of the user. In order to know the future location of the user, these two location-based services must use our system. The notion of vicinity is fixed by the location-based service with the parameter $\Delta r_{utility}$ in meters and the notion of future time with the parameter Δt_{future} in seconds. The parameters chosen for these two scenarios are summarized in Table 1.

We consider a utility of 1000 m for the public transportation scenario and 500 m for the taxi scenario. For sake of simplicity, we predefine several target time slices for each scenario: 10 for the public transportation scenario and 4 for the taxi scenario. As we previously mentioned in the evaluation of the location prediction component in Sect. 5.2, we decide to create two location trend models, one model with a time slice duration of 30 min, i.e., 48 time slices per day over one week and another model with a time slice duration of 1 h, i.e., 24 time slices per day over one week. Regarding the public transportation scenario and the first model, we select the 10 following time slices: in the morning from 7:00 am to 7:30 am, and at the end of the afternoon, i.e., from 5:30 pm to 6:00 pm, every working day. Regarding the taxi scenario and the first model, we select the 4 following time slices: Thursday from 11:30 pm to 0:00 am, Friday from 10:30 pm to 11:00 pm, Saturday from 4:30 pm to 5:00 pm and from 10:30 pm to 11:00 pm. Regarding the public transportation scenario and the second model, we select the 10 following time slices: in the morning from 7:00 am to 8:00 am, and at the end of the afternoon, i.e., from 6:00 pm to 7:00 pm, every working day. Regarding the taxi scenario and the second model, we selected the 4 following time slices: Thursday from 11:00 pm to 0:00 am, Friday from 10:00 pm to 11:00 pm, Saturday from 4:00 pm to 5:00 pm and from 10:00 pm to 11:00 pm. The number of prediction requests is equal to 100 for each scenario. The distribution of the prediction requests amongst the time slices for each scenario is done in a randomly distributed or in an evenly distributed manner. For example, this means that each target time slice of the first scenario is requested 10 times, i.e., 100 divided by 10, if the evenly distributed manner is chosen.

Metrics

We present three metrics below that enable to evaluate and compare the *utility/privacy tradeoff* LPPM of our system to the two other LPPMs.

Utility Metric. The first metric chosen for this evaluation aims at computing the utility in a binary manner. To do so, we extract one temporary predicted region $tmpRegion_{predicted}$ for a target time slice. We obfuscate this value by using a LLPM and obtain a new coordinates that are sent to a location-based service as the form of $loc_{predicted}$. We compute a reference zone $zone_{ref}$ that has a center equal to the center of $tmpRegion_{predicted}$ and a radius of $\Delta r_{utility}$. We also create a zone to check $zone_{toCheck}$ that has a center equal to the coordinates of $loc_{predicted}$ and a radius of $\Delta r_{utility}$. From these two zones, we compute if there is an intersection between the reference zone and the zone to check. If there is no intersection between these two zones, we count 0 for the utility and 1 otherwise, as detailed in Eq. 7. If the utility condition is satisfied, i.e., equal to 1, this means that the reliability of the information provided by the location-based service is ensured.

$$res_{utility} = \begin{cases} 0, & \text{if } zone_{ref} \cap zone_{toCheck} = \emptyset \\ 1, & otherwise \end{cases} \tag{7}$$

Location Privacy Metric (Uncertainty). The goal of the uncertainty metric is to evaluate if the location-based service, i.e., the adversary, is able to extract one single location by analyzing all the locations of a user received from our system for a specific time slice. In order to do so, we use a grid that discretizes space and helps to compute the density proportion $p_{density}$ of each visited cell of this grid. The density proportion $p_{density_i}$ is the number of predicted locations for a specific cell i of the grid out of the total number of predicted locations received a the target time slice. Each cell of the grid is a rectangle of approximately 100 m per 180 m on an average, i.e., a difference of 0.001 between two successive latitudes or longitudes. Equation 8 presents the uncertainty computation, in which i is the index of the ith visited cell by the user, n is the total number of visited cells by the user during the target time slice. A low entropy result means a low uncertainty of the adversary to highlight one single location, while a high entropy result means a high uncertainty.

$$res_{locationPrivacy_{unc}} = -\sum_{i=1}^{n} p_{density_i} \log_2 p_{density_i} \tag{8}$$

Location Privacy Metric (Accuracy). The goal of the accuracy metric is to compute the mean of all the distances between the center of the temporary predicted region $tmpRegion_{predicted}$ and all the locations $loc_{predicted}$ sent to the location-based service for a specific time slice. Equation 9 presents the accuracy computation, in which i is the index of the ith $loc_{predicted}$ and n is the total number of locations $loc_{predicted}$ sent to the location-based service for the target time slice. We also consider a function $distance$ that enables to compute the distance between two coordinates. The center of $tmpRegion_{predicted}$ is expressed with $tmpRegion_{predicted}.c$ in the equation. A low result means that the predicted locations are close to the raw predicted location provided by the location prediction component on an average.

$$res_{locationPrivacy_{acc}} = \frac{1}{n} \times \sum_{i=1}^{n} distance(tmpRegion_{predicted}.c, loc_{predicted_i}) \tag{9}$$

Table 2. Utility/location privacy results (48 time slices per day).

LPPM/Result	Utility result	Location privacy result (uncertainty)	Location privacy result (accuracy)
Utility/privacy tradeoff	**1.0**	**2.48**	**311.38**
Gaussian perturbation	0.50	2.46	16489.85
Grid-based rounding	0.58	0.0	4033.91

Table 3. Utility/location privacy results (24 time slices per day).

LPPM/Result	Utility result	Location privacy result (uncertainty)	Location privacy result (accuracy)
Utility/privacy tradeoff	**1.0**	**1.74**	**311.17**
Gaussian perturbation	0.50	1.73	16514.46
Grid-based rounding	0.60	0.0	3598.56

Results

First, we want to visually show the results of the application of the three LPPMs, which are the Gaussian perturbation, the rounding mechanism and the mechanism in our system that is the *utility/privacy tradeoff*. In Fig. 12, we can see the raw temporary visited region extracted from the location trend model in subfigure (a), the generation of 100 new locations with the Gaussian perturbation in subfigure (b), the generation of 100 new locations with the rounding mechanism in subfigure (c) and the generation of 100 new locations with the mechanism of our system in subfigure (d). Regarding the rounding mechanism, it is simply the example of another rounding mechanism that is not based on a grid but on a simple rounding to 2 decimals. As we can see in the subfigure (c), the creation of 100 new locations, rounded to 2 decimals, only generates the same location, which is exactly the same with the grid-based rounding mechanism for a specific size of cells. We can clearly see that the 100 new locations generated with the *utility/privacy tradeoff* mechanism with a parameter of 0.9 are closer to the center of the temporary predicted region.

Second, the results of the utility/privacy metrics are summarized in Tables 2 and 3 depending on the two time slice durations chosen to create the users' predictive models, i.e., 48 time slices and 24 time slices per day over one week. In these figures, we present two summaries of the three results corresponding to the three metrics: the utility, the uncertainty of the location privacy and the accuracy of the location privacy. The results of these three metrics for these two models were obtained as follows: we first extracted predicted temporary visited regions for all the target time slices for each scenario and for each user. Then, we created 100 new obfuscated coordinates from the predicted temporary visited region (i.e., raw value) of all the target time slices by using the three LPPMs and their parameters in a randomly and in an evenly distributed manners for each scenario and for each user. Finally, we computed the three metrics for all the target time slices of each LPPM and its related parameters and each scenario for each user. At the end of this evaluation process we decided to only compute the mean of each metric for all LPPMs and their parameters for all scenarios and for all users in order to obtain a summary view of the effects knowing that the parameters of the LPPMs were properly distributed. The goal of this evaluation with these three metrics was to maximize the utility in order to maintain a good quality of service, to maximize the uncertainty of an adversary when it performs a localization attack for a target time slice and to minimize the accuracy in order to also maintain a good quality of service. Regarding these three conditions, the summaries highlight that the *utility/privacy tradeoff* mechanism provides the best results. Even if the grid-based rounding has an uncertainty result of 0.0, the accuracy is not very high. We can also notice that the uncertainty result of the *utility/privacy tradeoff* mechanism and that of the Gaussian perturbation are very similar. However, their accuracy are completely different, i.e., low for the *utility/privacy tradeoff* mechanism and high for the Gaussian perturbation. Finally, regarding the utility, the result of the *utility/privacy tradeoff* mechanism outperforms those of the two other mechanisms.

6   Related Work

This section presents several works of the literature linked to the two main subjects of this paper. The first subject focuses on the works done in the field of the prediction of future locations of users. The second subject is linked to the existing LPPMs presented in the literature.

6.1   Location Prediction Models

There exist various models to predict future locations of users as detailed in the complete survey in [10]. In the literature, we can find different location predictive models related to different types of location prediction requests. For example, it is possible to predict future locations of users based on a time duration [11,15], to forecast the next location reached by a user [5,8,17], etc... Other papers aim at answering other location-based predictive requests, such as the prediction of the staying time in a particular visited cluster or when the user will reach or leave a visited cluster [8], the prediction of the number of users that will enter in a specific zone [4] and much more. Regarding the prediction of the next visited location of a user, the models are often based on frequently visited clusters of users, such as presented by Gambs et al. in [7]. For instance, Etter et al. develop several predictors and propose different blending strategies in order to maximize the accuracy of the prediction of the next location visited by a user. Other remaining works focus on the performance on new or existing predictive models. In [16], Xu et al. present a way to prune an order-k Markov chain model in order to efficiently compute long-term predictive range queries.

In this paper, the location prediction goal is to extract a future location of a user from her predictive model and based on a time duration from the current time. In the literature, it has been shown that some predictive models can work better for near location predictions and others are more suited for distant

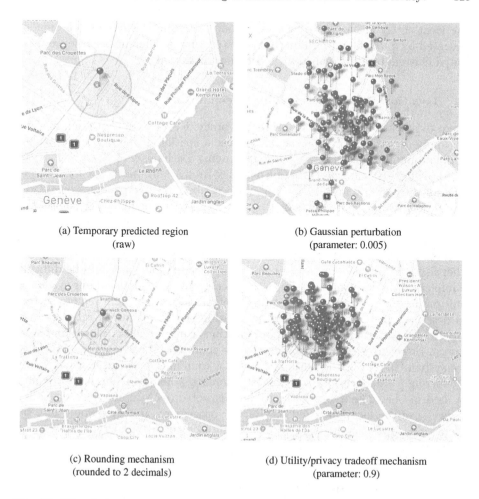

(a) Temporary predicted region
(raw)

(b) Gaussian perturbation
(parameter: 0.005)

(c) Rounding mechanism
(rounded to 2 decimals)

(d) Utility/privacy tradeoff mechanism
(parameter: 0.9)

Fig. 12. Visual description of the impact of the different LPPMs on a temporary predicted region.

location predictions. In [11], Jeung et al. present a hybrid prediction model for moving objects. Their model uses motion functions for near location predictions, while their model extracts predicted locations based on trajectory patterns for distant location predictions. The structure in which they store the trajectory patterns of a user is a trajectory pattern tree. Their predictive model is close to our location trend model because they used several portion of times in order to structure it. Their model takes into account sequences of spatial clusters to fill it, which is not the case in our predictive model. Sadilek and Krumm present a model that enables to predict long-term human mobility in [15] up to several days in the future. Their method uses a projected eigendays model created by analyzing the periodicity of the mobility of a user as well as additional features.

The location trend model we propose in our system is also close to the model presented by Sadilek and Krumm but the structure is different because we used visited clusters to fill it (called visited regions in Sect. 4).

6.2 Location Privacy Preserving Mechanisms

There exist various LPPMs to protect a predicted location. For example, we can apply a spatial perturbation [1,2,6], use a spatial cloaking mechanism [9], sending dummy locations [12] or use a rounding mechanism [1,13].

Applying a spatial perturbation enables to spatially modify a location as mentioned by several authors in [2,6]. As described in these papers, we can add spatial noise to the coordinates of a location. However, the higher the noise, the higher the utility of the location-based service decreases in our context. Regarding the spatial cloaking presented by Gruteser and Grunwald in [9], the predicted location is only sent if the user cannot be distinguishable from at least $k-1$ other users, i.e., considered as *k-anonymous*. This mechanism would be difficult to implement in our context because the user's predictive model of our system is built on her mobile device and not in a centralized manner on a remote server. As described in [12], sending dummy locations is interesting in order to add noise if and only if multiple predicted locations can be sent to a location-based service. However, this is not possible in the context of the predictive request of our system. In [1,13], the authors present a rounding mechanism, which aims at blurring a raw location to a nearest reference point. This mechanism could be used in our context but without the guarantee to maintain a good quality of service of the location-based service. Cryptography strategies would also be interesting to protect the location privacy in users as mentioned in [10] but not in the context of our work.

7   Conclusion and Future Work

In this paper, we presented a system that provides future locations of users to location-based services. The key elements of this system is that it will ensure a good quality of service while preserving the location privacy of the user. It contains two components: a location prediction component and a location privacy component. The first component updates the *location trend model* of the user and extracts temporary predicted regions from the model. The second component aims at blurring the temporary predicted regions with a *utility/privacy tradeoff* mechanism. The user is a key actor of the system because she can indicate the temporal and spatial percentages she is willing to sacrifice in order to protect her privacy. Although the temporal percentage helps to protect the location privacy of the user, this parameter is only used during the prediction, i.e., in the location prediction component. Regarding the spatial percentage, this parameter is only used in the location privacy component. We evaluated the two components of the system with a location prediction analysis for the first component and with a utility/location privacy analysis for the second component. The results obtained for the *location trend model* are satisfactory and can exceed 70% in terms of prediction accuracy. In addition, the *utility/privacy tradeoff* mechanism provides the best results compared to two other LPPMs. This research could open one future work that focuses on the implementation of the entire system on real mobile devices in order to assess the performance of the entire system. This future work could help to improve the updates of the *location trend model*, the efficiency of the extraction of the temporary predicted regions and finally the process of the *utility/privacy tradeoff* mechanism.

References

1. Agrawal, R., Srikant, R.: Privacy-preserving data mining. In: ACM SIGMOD Record, vol. 29, pp. 439–450. ACM (2000)
2. Armstrong, M.P., Rushton, G., Zimmerman, D.L.: Geographically masking health data to preserve confidentiality. Stat. Med. **18**, 497–525 (1999)
3. Backes, M., Humbert, M., Pang, J., Zhang, Y.: walk2friends: inferring social links from mobility profiles. In: Proceedings of the 2017 ACM SIGSAC Conference on Computer and Communications Security, pp. 1943–1957. ACM (2017)
4. Chapuis, B., Moro, A., Kulkarni, V., Garbinato, B.: Capturing complex behaviour for predicting distant future trajectories. In: Proceedings of the 5th ACM SIGSPATIAL International Workshop on Mobile Geographic Information Systems, pp. 64–73. ACM (2016)
5. Gambs, S., Killijian, M.O., Nuñez Del Prado Cortez, M.: Next place prediction using mobility Markov chains. In: MPM - EuroSys 2012 Workshop on Measurement, Privacy, and Mobility - 2012, Bern, Switzerland, April 2012. https://hal.inria.fr/hal-00736947
6. Gambs, S., Killijian, M.O., Núñez del Prado Cortez, M.: Show me how you move and i will tell you who you are. Trans. Data Privacy **4**(2), 103–126 (2011). http://dl.acm.org/citation.cfm?id=2019316.2019320
7. Gambs, S., Killijian, M.O., del Prado Cortez, M.N.: Next place prediction using mobility Markov chains. In: Proceedings of the First Workshop on Measurement, Privacy, and Mobility, p. 3. ACM (2012)
8. Gidófalvi, G., Dong, F.: When and where next: individual mobility prediction. In: Proceedings of the First ACM SIGSPATIAL International Workshop on Mobile Geographic Information Systems, pp. 57–64. ACM (2012)
9. Gruteser, M., Grunwald, D.: Anonymous usage of location-based services through spatial and temporal cloaking. In: Proceedings of the 1st International Conference on Mobile Systems, Applications and Services, pp. 31–42. ACM (2003)
10. Hendawi, A.M., Mokbel, M.F.: Predictive spatio-temporal queries: a comprehensive survey and future directions. In: Proceedings of the First ACM SIGSPATIAL International Workshop on Mobile Geographic Information Systems, MobiGIS 2012, pp. 97–104. ACM, New York (2012). https://doi.org/10.1145/2442810.2442828. http://doi.acm.org/10.1145/2442810.2442828
11. Jeung, H., Liu, Q., Shen, H.T., Zhou, X.: A hybrid prediction model for moving objects. In: IEEE 24th International Conference on Data Engineering, 2008. ICDE 2008, pp. 70–79. IEEE (2008)
12. Kido, H., Yanagisawa, Y., Satoh, T.: An anonymous communication technique using dummies for location-based services. In: Proceedings of the International Conference on Pervasive Services 2005, ICPS 2005, Santorini, Greece, 11–14 July 2005, pp. 88–97 (2005). https://doi.org/10.1109/PERSER.2005.1506394

13. Krumm, J.: Inference attacks on location tracks. In: LaMarca, A., Langheinrich, M., Truong, K.N. (eds.) Pervasive 2007. LNCS, vol. 4480, pp. 127–143. Springer, Heidelberg (2007). https://doi.org/10.1007/978-3-540-72037-9_8. http://dl.acm.org/citation.cfm?id=1758156.1758167

14. Moro, A., Garbinato, B.: Respred: a privacy preserving location prediction system ensuring location-based service utility. In: Proceedings of the 4th International Conference on Geographical Information Systems Theory, Applications and Management, GISTAM, vol. 1, pp. 107–118. INSTICC, SciTePress (2018). https://doi.org/10.5220/0006710201070118

15. Sadilek, A., Krumm, J.: Far out: predicting long-term human mobility. In: AAAI (2012)

16. Xu, X., Xiong, L., Sunderam, V., Xiao, Y.: A Markov chain based pruning method for predictive range queries. In: Proceedings of the 24th ACM SIGSPATIAL International Conference on Advances in Geographic Information Systems, GIS 2016, pp. 16:1–16:10. ACM, New York (2016). https://doi.org/10.1145/2996913.2996922. http://doi.acm.org/10.1145/2996913.2996922

17. Ying, J.J.C., Lee, W.C., Weng, T.C., Tseng, V.S.: Semantic trajectory mining for location prediction. In: Proceedings of the 19th ACM SIGSPATIAL International Conference on Advances in Geographic Information Systems, pp. 34–43. ACM (2011)

18. Zang, H., Bolot, J.: Anonymization of location data does not work: a large-scale measurement study. In: Proceedings of the 17th Annual International Conference on Mobile Computing and Networking, MobiCom 2011, pp. 145–156. ACM, New York (2011). https://doi.org/10.1145/2030613.2030630. http://doi.acm.org/10.1145/2030613.2030630

Enabling Standard Geospatial Capabilities in Spark for the Efficient Processing of Geospatial Big Data

Jonathan Engélinus, Thierry Badard[(✉)], and Éveline Bernier

Centre for Research in Geomatics (CRG), Laval University, Quebec, Canada
jonathan.engelinus.1@ulaval.ca,
thierry.badard@scg.ulaval.ca,
eveline.bernier@crg.ulaval.ca

Abstract. Nowadays, big data are in the midst of many scientific, economic and societal issues. While most of these data include a spatial component, very few big data processing systems are able to manage this particular component. The authors have assessed the capabilities and limits of current solutions and have concluded that most of them are neither efficient nor extensive enough for spatial data. Furthermore, none of them fully complies with ISO standards and OGC specifications in terms of spatial processing. The authors have sought a way to overcome these limitations and have defined a system in greater accordance with the ISO-19125 standard. The proposed solution, called Elcano, is an extension of Spark complying with ISO-19125, allowing the SQL querying of spatial data and including an original spatial indexation system. Tests demonstrate that Elcano surpasses current available solutions on the market.

Keywords: Big data · ISO-19125 · Spatial indexation · Elcano · Magellan spatial spark · Geospark · Geomesa · Simba · Spark SQL

1 Introduction

The need for the management and processing of large data volumes is gaining increasing importance for industry and governmental bodies. For instance, Walmart has to process more than 2.4 petabytes of data per day and Google ten times as much. This also applies in many research domains, such as epidemy prevention [18], decision making in management [34] and e-commerce optimization [2]. According to IDC (International Data Corporation, https://www.idc.com), this situation is not about to stop as the total volume of data increases by 90% every two years [23].

Most notably, the convergence between internet and cartography [35] and the proliferation of sensors (e.g. GPS in smartphones, accelerometer, camera, …) have contributed to the significant increase of spatial data production and the need for their analysis [4]. This situation complicates the cartographic process as it becomes difficult to manage and represent such large quantities of data in a conventional way [16]. MapReduce [10] is an algorithm that allows the parallel and distributed processing of data for a faster execution. The data to process is also distributed among the servers by

L. Ragia et al. (Eds.): GISTAM 2018, CCIS 1061, pp. 133–148, 2019.
https://doi.org/10.1007/978-3-030-29948-4_7

the Hadoop Distributed File System (HDFS), which is provided by default with Hadoop. The result is a high degree of horizontal scalability, which can be defined as the ability to linearly increase the performances of a multi-server system to meet the user's requirements in terms of processing time.

The Hadoop environment [44], is built around this algorithm. It is currently one of the most important projects of the Apache Foundation and a de facto standard for Big Data processing and management [19]. In addition, an ecosystem of interoperable elements has been built up around Hadoop, that enables the management of other aspects, such as streaming (e.g. Storm), serialization (e.g. Avro) and data analysis (e.g. Hive).

In 2014, the University of Berkeley's AMPLab has started working on an interesting alternative to MapReduce and HDFS. Named Spark (http://spark.apache.org/), this new element of the Hadoop ecosystem distributes data and processing code on small blocks called RDD ("Resilient Distributed Dataset") through the whole cluster RAM. This architectural choice limits hard drive accesses and is up to ten times faster than conventional MapReduce and Hadoop uses [48, 49], although this comes at the cost of a greater RAM load [25].

With an ever-growing amount of geospatial data available, it would be profitable to use these capabilities in order to analyze the spatial component which, according to Franklin, is present in 80% of all business data [21]. Moreover, it has been stated that a better use of Big data spatial aspect could grant $100 billion to services providers and about $700 billion to end-users [33]. Furthermore, spatial Big data management finds itself in the midst of many important economic, scientific and societal issues. Querying spatial Big data in a SQL syntax could empower analysts with new analytical capabilities, provide a better understanding of a phenomenon's spatial distribution and bring new insights, while using a query language they are already familiar with.

A part of Spark called Spark SQL [3] addresses precisely this need. It supplements Spark RDD with a new level called "DataFrames". These additional structures organize data into temporary tables that can be queried using SQL. Spark SQL also optimizes Spark processes, thanks to strategic queries auto-rewriting and data serialization. It is a flexible solution, allowing the definition of two personalized structures: "User Defined Type" (UDT) and "User Defined Function" (UDF) for additional kinds of data and processing. Spark SQL also appears to be an interesting solution in terms of performance, as it processes data at least 7 times faster than Impala, another Hadoop element supporting SQL [46].

However, Spark doesn't provide any spatial management options. The build of a specific system on top of it to manage and handle this specific information is thus required. To guarantee interoperability, this system has to comply with existing spatial standards like ISO-19125. To be efficient, it has to index spatial data in such a way they will be retrieved efficiently and rapidly. But how to do that in the context of Big Data, where spatial data are fragmented between different servers? Before we address that complex question, we will consider several cur-rent Hadoop and Spark systems that support spatial data management.

The Hadoop environment proposes few systems that support spatial data management. For example, Hadoop GIS [1], Geomesa [27], Pigeon [12] and Spatial Hadoop [11], although all are more prototypes than mature technologies [4]. In addition, most of them only rely on Hadoop's core version and its MapReduce algorithm without fully scaling the RAM processing power, like Spark.

Only five Hadoop-based systems pro-pose a spatial data management relying on Spark. The first two, Spatial Spark [46] and GeoSpark [48] have been developed upon the Spark core and they do not include SQL extension. Hence, with the current Spatial Spark version, we can only interact with data using command lines instead of SQL queries. GeoSpark only uses its own spatial extension of the Spark RDD type, which does not directly comply with Spark SQL [47]. The third system, Magellan [39], defines spatial data types directly available in Spark SQL, but does not correctly manage some spatial operations like the union of disjointed polygons, symmetric differences involving more than a geometry and the creation of an envelope. Furthermore, Magellan is the only one among the five Spark prototypes with no spatial indexation management. The fourth, Simba [45], only enables SQL data querying for points geometries, and doesn't allow the possibility to trigger standard spatial functions. At last, the fifth prototype is the Spark's Geomesa extension. This extension is quite limited regarding spatial operations as it can only search for points that are included in envelopes. Furthermore, its performances are limited [45] compared to other solutions, which may be explained by the fact that it relies on a key storage technology (Accumulo, https://accumulo.apache.org/).

We are finally able to conclude that there is currently no geospatial data management system that fully manages all kinds of 2D geometry data types and enables their efficient and actionable SQL querying. Each model implemented in the five studied prototypes presents limited capacities regarding the types of geometry they support and the spatial processing and indexation functions they offer.

2 Limits in the Geospatial Capabilities Supported by Current Solutions

In order to better assess the capabilities of the current systems to fully manage the 2D spatial component, we can use the ISO-19125 standard as a guideline. This two-part standard describes the 2D geometry types and the geospatial functions and operators [28] and their expression in the SQL language [29] that a system must implement to basically store 2D geospatial data and support their querying and analysis in an interoperable way. In this context, we will first intro-duce the geometry types supported by the different systems. Then, we will analyze which spatial functions they offer and whether they can be extended to easily implement the missing ones. Finally, we will study how they manage the spatial indexation issue, which is crucial when dealing with geospatial data.

2.1 Geometry Data Types

A system complying with ISO 19125-1 is supposed to handle the seven main 2D geometry types that can be built by linear interpolation. These geometries can be divided into three simple types (point, polyline and polygon) and into four composite types (multipoint, multipolyline, multipolygon and geometry collection). Here is a study of how the current systems meet this standard.

Spatial Spark and GeoSpark integrate all these types of geometries because their model relies on the JTS ("Java Topology Suite", https://www.locationtech.org/proposals/jts-topology-suite) library, which has been designed to meet ISO standards and OGC recommendations [9]. Geomesa also sup-port all the geometries in its current version [7], while Simba only support points geometries.

The case of the HortonWorks Magellan system is less simple. It enables the processing of points, polylines and polygons. This may seem sufficient if we assume, as it is the case of the system's designers [40], that compound geometries are reducible to tables of geometries. However, such an approach can only lead to a dysfunctional system. Indeed, by not being able to explicitly create actual complex geometries, such arrays are not allowed as operands of a spatial function and their re-turn as a result of a spatial operation like the union of disjoint polygons causes a type error. Magellan's limitations are also due to the use of ESRI Tools as a spatial library, which cannot process all the 2D geometry types defined by the ISO-19125 standard. ESRI Tools do not support the collection type, while the multi-polygon type is only partially implemented. Furthermore, the adaptation of WKT ("Well-Known Text") pro-vided by ESRI Tools does not comply with the ISO standards and the OGC recommendations. Table 1 summarizes the limitations of the different solutions regarding the ISO-19125-1 requirements. Those related to ESRI Tools have been added to give an idea of the limits that they involve on the evolution of Magellan.

Table 1. Coverage of the different 2D geometry types specified by ISO-19125 in studied prototypes [15].

	Geo Spark	Spatial Spark	Simba	Geomssa	Magellan	ESRI
Point	Yes	Yes	Yes	Yes	Yes	Yes
Polyline	Yes	Yes	No	Yes	Yes	Yes
Polygon	Yes	Yes	No	Yes	Yes	Yes
Multi-point	Yes	Yes	No	Yes	No	Yes
Multi-polyline	Yes	No	Yes	No	Yes	Yes
Multi-polygon	Yes	Yes	No	Yes	No	In part
Collection	Yes	Yes	No	Yes	No	No

2.2 Spatial Functions and Operators

ISO 19125-2 [29] specifies which spatial functions (relations, operations, metric functions and methods), a spatial data management system should implement in SQL to comply with the ISO 19125-1 standard. It does not specify the way these methods have to be implemented. It only defines their signatures. These functions define the minimal set of operations a system must implement to enable basic and advanced spatial analysis capabilities. Even if these functions have been defined for querying data in classic spatial DBMS, their usage in geospatial Big data management systems still pertain. Nevertheless, the application of the ISO-19125-2 standard re-quires a system that support SQL queries and personalized SQL functions. This section details how the five studied systems partly implement the standard and describes their extension capabilities.

As mentioned earlier, Spatial Spark only uses the Spark's core functions. So, it supports RDD but not the DataFrames or SQL queries. Accordingly, the application of the ISO 19125-2 standard to Spatial Spark seems impossible without a full reimplementation.

GeoSpark extends the RDD type of Spark and is therefore not directly compliant with Spark SQL. Nevertheless, according to [47], the integration of such a capability is planned for a future release that should provide a way to change these RDD's into DataFrames. But neither does he describe a general process for it, nor how to apply SQL queries afterwards. Therefore, the current version of GeoSpark does not seem to be compliant with the ISO 19125-2 standard because all geometry types cannot be managed from SQL queries.

Simba released its own adaptation of Spark SQL, which might, in theory, enable the use of SQL queries and the creation of User Defined Functions. In practice however, the only accessible geometry is the point. Furthermore, the syntactic analyzer does not always work properly. By example, it forces to write "IN" before "POINT(x, y)" even without a context of inclusion. Simba is therefore not a mature and reliable solution that could meet the ISO 19125-2 standard.

Until recently, Geomesa's Spark extension only used Spark's core. But a recent version tries to integrate with Spark SQL. However, this solution remains restrained by the mandatory use of the CQL format and the Accumulo database [7]. As a result, Geomesa does not allow a transparent and technology agnostic implementation of ISO-19125-2.

Magellan does not directly manage SQL either. But it defines User Data Types for the point geometries. It is therefore tempting to assume that the addition of User Defined Functions to its model should be enough to allow the SQL functions of the ISO-19215-2 standard. In practice however, the extension of Magellan with these functions only covers two thirds of spatial relations, half of the spatial operations and a small part of spatial methods specified by the ISO-19125-2 standard. These limitations are due to both implementation errors and the choice of the ESRI Tools library, which only partially meets the ISO-19125-2 standard.

Therefore, none of the studied systems, in their current states, totally comply with the ISO-19125 standard.

2.3 Spatial Indexation Management

In general, "data indexation" could refer to transformations applied by a system to its acquired data, with the consequence of an acceleration of their reading or writing. When data have a spatial component, specific methods should be used. They rely on the positions of all data and their relations of proximity. Indeed, spatial data have strong interconnections that can be used for their reading. "Spatial data indexation" can thus be defined as the reorganization of spatial data, typically by using their proximity relations, with the purpose of accelerating their processing [13]. The intended result is a significant acceleration of high-cost operations (like spatial jointure, which without indexation might require a cartesian product between the data).

There are many spatial indexation technics that we can classify in two categories: plane and hierarchical [13]. In the first approach, the territory is simply divided by sectors and the data are classified by sector. The second approach builds a tree with data classified by their proximity relations, for a finer but less homogenous classification. Figure 1 compares the Grid-file plane indexation method [37] with the Quadtree hierarchical indexation method [20]. Gridfree simply uses a regular grid for data classification. Quadtree recursively divides each sector containing data (and only them) in four subsections.

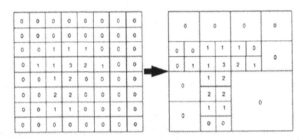

Fig. 1. Grid-file (in left) versus Quadtree (in right) [30].

As illustrated, the hierarchic indexation is the most precise and the most economical in terms of created sectors. But a hierarchic spatial index is also more complex to build and seems more difficult to implement in a scalable way because of the sectors' irregular repartition. There exists also one hybrid approach [41] combining the R-Tree hierarchical indexing method [26] and the Geo-Hash plane indexation [36]. But the question of its implementation in a Big data context using Hadoop remains partly open.

Four of the studied systems provide a spatial indexation component, but none of them is efficient and extensible.

The spatial indexation component of Spatial Spark directly uses the hierarchical and plane methods of the JTS spatial library, which is not conceived for Big data processing in a multi-server environment. GeoSpark proposes a more integrated and efficient spatial indexation module [47], but that cannot be managed using SQL queries. The indexation component of Simba is described as more efficient by its developers [45], but has important limitations and bugs that we have already covered in this paper. Finally, Geomesa offers poor performances because it relies on a specific database system [45], which drastically increases the processing time.

2.4   Synthesis of Limitations

Table 2 sums up the main limitations of the studied systems. It first recalls their most problematic limitations. Then it reminds the geometry types they support and as a result their degree of conformance to the ISO 19125 standard. Next, it indicates whether they manage SQL and whether they comply with the ISO-19125-2 standard.

Table 2. Limitations of current spatial Big data processing systems [15].

	Magellan	Spatial Spark GeoSpark	Simba	Geomesa
Main limitation	Use a limited spatial library	Inextensible to SQL	Syntactic bugs, no extensible	Force to use a NoSQL DB
Types of geometries	Only simples	All	Only point	All
ISO-19125-1	In part	Yes	In part	Yes
ISO-19125-1	In part	Yes	In part	Yes
SQL management	No, but extensible	No	Yes (replace Spark SQL)	Yes, limited by CQL
ISO-19125-2	In part (by extension)	No	No	In part
Spatial indexation	No	Yes, but not efficient or not extensible	No	No

Next section presents a new system de-signed for the rapid processing of geospatial Big data (vector data only) and their efficient and interoperable management. It relies on Spark and overcomes limitations previously identified in current state-of-the-art solutions. This prototype is named Elcano. Its release as an open source project has not yet been performed but it is envisaged.

3   Presentation of ELCANO

The main objective leading the design of Elcano is to model a spatial Big data processing and management system that surpasses the other systems studied here, not only in terms of performance but also in terms of capabilities offered to deal with spatial data. To do so, it must: (1) integrate all the 2D geometry types defined in the ISO-19125 standard; (2) enable the use of associated spatial functions; (3) implement all spatial relations, operations and methods also defined by the ISO-19125-2. For example, a call to the SQL function ST_Intersects has to indicate if two generic geometry objects intersect or not. The system must also provide functions to load spatial data in a simple and generic way, so it can easily be fed by different spatial data formats. It must ensure in-memory data persistence for a faster processing of the geospatial component. It has to be easily extensible to potentially support new geometry types or extensions to the geometry types defined in the ISO-19125 standard (e.g. the inclusion of elevation in geometric features definition, i.e. 2.5D data). Finally, it must offer good processing performances in comparison with current processing systems. A model seeking to meet these objectives is presented and justified below.

3.1   Architecture

Figure 2 illustrates the model on which Elcano is based. The Geometry package implements the geometries and the spatial functions of Elcano. The Loader package enables the
use of SQL spatial functions. Data persistence for data processing and retrieval is managed by the Table class and by the conversion methods included in the GeometryFactory
class. Finally, the Index package indexes spatial data for faster processing. Next section
presents in more detail the implementation of these different capabilities.

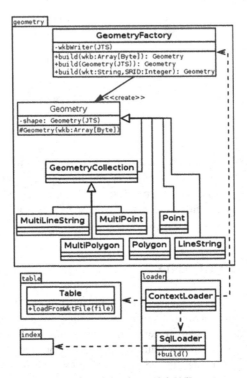

Fig. 2. Elcano's model [15].

2D Geometry Types Management. The geometry package of Elcano contains a
concrete class for each geometry type de-scribed in the ISO 19125 standard. These
classes use the JTS spatial library, which has specifically been built to comply with
many ISO standards (including ISO 19125) and the OGC recommendations [9]. By
using this library, we avoid the problems faced by Magellan and previously described.
Also, we could have chosen to use JTS classes directly, such as Spatial Spark and
GeoSpark, but for optimization purposes, we have decided to not be constrained by the
implementation of a chosen spatial library. To this end, the Elcano geometry package
uses a JTS-independent class hierarchy by applying the "proxy" design pattern [22].
This choice of design allows to optimize, whenever possible, the JTS methods by
overwriting them.

Spatial SQL Functions Management. In order to make the spatial functions and operators defined in ISO 19125 available as SQL functions in Elcano, different User Defined Functions (UDF) have been de-fined. All these functions are in fact shortcuts to the different methods supported by the geometry types (i.e. classes included in the geometry package) and specified in the ISO 191125 standard. The build() method of the SqlLoader class in the Loader package is in charge of declaring all these functions at the initialization stage of the application.

Spatial Data Persistence. Elcano provides a unified procedure for the loading of all 2D geometry types and their persistence. The Table class enables the definition of geometric features in WKT. WKT is a concise textual format defined in the ISO 19125 standard. Elcano thus allows to load tabular data (e.g. from a CSV file where the geometry component of each row is defined in WKT) in the form of an SQL temporary table. The management of more specific formats like JSON [6], GeoJSON or possible spatial extensions to Big data specific file formats like Parquet [43] can also be easily added to the system by simply inheriting the Table class.

Data Types Extensibility. The GeometryFactory class implements the "abstract factory" design pattern [42] and allows the extensibility of Elcano. Other geometry types than those defined in the ISO-19215 standard could eventually be added, such as Triangles and TINs to support DTMs (Digital Terrain Models).

Spatial Indexation. The Index package contains the classes in charge of the spatial indexation of data stored in Elcano. It drastically speeds up all spatial processes. Figure 3 details the model of this part of Elcano.

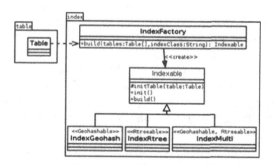

Fig. 3. Elcano's indexation model.

This component is inspired from the one implemented in Spatial Spark, but some hooks have been added for better performance.

The resulting spatial indexation system provides three indexation methods. The first two are the GeoHash (plane indexation) and the R-Tree (hierarchical indexation) methods. The conceptual reasons underlying this choice will be presented in a future paper dedicated to spatial indexation. A hybrid indexation method combining the two and inspired from [41] is also proposed. Our tests reveal a good complementarity between these three indexation methods: GeoHash is better for indexing data with large geometries and inhomogeneous repartition (such as large lakes in a country), R Tree is better for homogeneous data with small geometries (such as houses in a city), and the hybrid index is better for intermediary cases. The class model uses the "abstract factory" design pattern, which allows the implementation of other indexes if needed.

The benchmarks section explores the performance of the spatial index mechanism implemented in Elcano.

4 Benchmarks

The present section compares the performances of Elcano and Spatial Spark. The latter has been chosen because it is the only one among the other systems presented in this paper that can be extended to support SQL queries (though it requires a significant reimplementation). Both systems are also compared with a well-known and widely used classical spatial database management system: PostGIS [38]. From our point of view, PostGIS is a reference implementation of the ISO 19125 standard. In addition, it proposes efficient and reliable spatial indexation methods.

For the needs of this benchmark, Elcano and Spatial Spark have been in-stalled on a cluster of servers using a master server with 8 Gb of RAM and nine slave servers with 4 Gb of RAM. Each of these computers uses the CentOS 6.5 operating systems and height Intel Xeon 2.33 GHz processors. PostGIS has been optimized with the pgTune library (https://github.com/le0pard/pgtune) and tested in comparable conditions.

In each of the 3 tests performed, we count the number of resulting elements from a spatial join between two tables. We group the elements of these tables by pair, according to a given spatial relation, namely the intersection. This spatial relation has been chosen because it implies complex and sometimes time-consuming processing. The use of a fast and reliable spatial indexation system is also of importance in such a process. The contents of the tables used in the test is fixed. The management of changing data is out of the scope of the tests.

Test 1 compares the execution time of the three systems while increasing the volume of data. It consists in counting the intersections between an envelope around the province of Quebec and seven sets of points randomly dispatched in an envelope around the Canada. These seven sets contain respectively 1000, 10 000, 100 000, 1 million, 100 million and one billion points.

Table 3 presents a synthesis of the first test results. In order to facilitate their comparison, the duration cumulates the indexation time and the first query time. It appears that beyond one million points, Elcano's performances are better than those of Post-GIS and Spatial Spark. PostGIS is the best choice for lower volumes but encounter a significant slowdown after a certain thresh-old: it requires many hours to process 100 million points against five minutes for Elcano. The difference between Spatial Spark and Elcano is more tenuous but in-creases in favor of Elcano as data volume increases.

The drop in PostGIS performances when data volume increases can be explained by its weak horizontal scalability: this system is not designed for Big data management. In return, performances of Elcano when compared to Spatial Spark can be explained by its usage of Spark SQL. Indeed, the latter uses specific query optimizations and Spark's caching system [3]. But for small data volumes (under one million points), the classical

PostGIS solution is better, probably because of its simpler distributed treatments architecture. In a similar way, the best performances of Spatial Spark between one and ten million points can be explained by the additional treatments imposed by the use of Spark SQL in Elcano.

Test 2 compares the horizontal scalability of Elcano and Spatial Spark from one to nine servers. It compares the count of the intersections between the envelope of the province of Quebec and one billion points randomly dispatched in a bounding box around Canada. PostGIS performance is not measured for this test because the previous clearly underlies its poor performances for large data volumes and there is no way to distribute the processing between many servers as PostgreSQL has not been de-signed for horizontal scalability.

Table 3. Test 1 – Processing time when increasing data volume [15].

Volume (points)	PostGIS (ms)	Spatial Spark (ms)	Elcano (ms)
1000	234	6 543	9 516
10 000	326	6 622	9 714
100 000	3 783	8 301	9 030
1 000 000	29 898	8 301	10 747
10 000 000	269 257	20 487	17 099
100 000 000	5 752 821	55 017	37 378
1 000 000 000	More than 10 h	399 100	273 074

The Table 4 presents a synthesis of the results for this second test. Spatial Spark and Elcano both have a good horizontal scalability. Furthermore, the execution time of the two systems presents a similar drop from one to nine servers: 87,4% for Spatial Spark and 87,2% for Elcano. But Elcano remains globally 1.5 faster than Spatial Spark regardless of the number of servers.

Elcano's superior speed in this second test can be explained by its usage of Spark SQL. Otherwise, the rates of scalability of the two systems are very close, maybe because both rely on the JTS spatial library for the implementation of the spatial analysis algorithms.

Table 4. Horizontal scalability when the number of servers increases (15).

Servers	Spatial Spark (ms)	Elcano (ms)
1	3 349 414	2 196 344
2	1 718 672	1 123 153
3	1 143 790	762 536
4	875 284	588 401
5	696 195	473 635
6	586 211	391 297
7	511 111	340 784
8	456 446	314 796
9	423 647	280 761

Test 3 compares more finely the performances of PostGIS, Spatial Spark and Elcano. It counts the intersections between one million points in an envelope of Canada and the points in a copy of this set. Therefore, a total of 100 billion intersection tests (spatial join) are processed. The execution time is spread between indexation time, first query time (cold start) and second query time (hot start). Hot start queries are more representative of the response times in a running environment in production. Indeed, while indexing is only necessary once for the two given tables, an SQL query must be started for each spatial join operation applied to them.

Table 5 offers a summary of the results for this third test. PostGIS presents a spatial indexation time a bit shorter than Spatial Spark, but the execution time of its first SQL query is then much longer. Elcano presents the best performances in all cases: its indexation time is 5 times lower than PostGIS and the execution of its first query is 2 times faster than Spatial Spark. Elcano is also the only solution to execute a second SQL query on the same data significantly faster than the first: the second execution is 11.3 times faster in R-Tree indexation mode and 5 times faster in GeoHash mode, which can surely be explained by Spark SQL's caching system. It also appeared that the indexation in R-Tree mode is longer than the indexation with GeoHash but gives better results for the second query. This may be due to the lower complexity of plane compared to hierarchical indexation.

Table 5. Test 3 – execution time is spread between indexation time, first query time and second query time. Inspired by [15].

Solution	Indexation time (ms)	First query (ms)	Second query (ms)
PostGIS	29 756	100 742	100 742
Spatial Spark	36 824	36 824	36 824
Elcano (R-Tree mode)	13 578	15 754	1 393
Elcano (GeoHash mode)	5 518	13 492	6 862

To sum up, above a given data volume, Elcano surpasses PostGIS and Spatial Spark in terms of execution speed. It presents a scalability similar to Spatial Spark, but a better execution time when the number of servers increases.

5   Conclusion and Perspectives

While Big data including a spatial component are in the midst of many current scientific, economic and societal issues, and in spite of the fact that Hadoop has become a mature de facto standard, the number of processing and management systems for spatial Big data using the Hadoop environment remains limited. Also, the available solutions are prototypes with limited capabilities. Moreover, while Spark is surely the most promising Hadoop module for this type of processing, only five solutions are based on Spark and none of those can entirely handle all geometry types. Finally, SQL spatial functions specified in the ISO 19125 standard are either missing or not compatible with an efficient spatial indexation in these solutions.

To tackle these issues, we proposed a new spatial Big data processing and management system relying on Spark. This system called Elcano is based on the SQL library of Spark and uses the JTS spatial library for its compliance with the ISO's standards. Thanks to this approach, all spatial SQL functions and operators defined by the ISO 19125 standard are fully supported.

The proposed model on which Elcano relies is not a simple implementation of JTS. It comes with the possibility to use SQL spatial queries with a data model that can evolve. Furthermore, it integrates the geometric data types on the Big data con-text and comes with a scalable spatial indexation system which will be detailed in an upcoming paper.

In addition, Elcano offers better performances than Spatial Spark and a similar scalability. Indeed, Elcano comes with an extensible and efficient spatial indexation system, which allow its adaptation to several types of spatial Big data. Many of the studied Spark-based solutions tried to provide a spatial indexation system, but either without complete efficiency (Spatial Spark, Accumulo) or with too important rewritings of the Spark Core (Simba, GeoSpark) coupled with the impossibility of directly using Spark SQL. Elcano comes with an alternative which is more respectful of Spark and Spark's SQL structures. A detailed study of all the possibilities in term of spatial indexation management remains however to be done. A way to address it could be to adapt the no Hadoop solution defined by [8] to the Spark environment, but there are also many classical spatial data indexation methods that could be explored and adapted for a Big data context. A general reflection on what could be gained from recent mathematical works for the optimization of spatial indexation also lacks in previous works.

In a larger perspective, it could be interesting in a near future to enable the management of the elevation together with dedicated data types such as Triangles and TINs in the current model. Raster data types are also considered for inclusion. The use of RasterWKT could be considered to achieve such a support. That would give us the flexibility to apply the model to many new challenging situations such as the processing of large collection of images coupled with vector data analytics capabilities or the building and analysis of high-resolution digital elevation models (DEM) or DTM without being compelled to split them into tiles hence allows for the processing of very huge amount of spatial data.

The current version of Elcano only support batch data processing, but it could be very interesting to add the possibility of processing and displaying continuously received data, i.e. in streaming [14]. Such an extension could indeed enable the design of real time Big data geospatial analytical tools that will help users (analysts, decision makers, …) to make fully informed decisions on up-to-date data and in a shorter period of time. Furthermore, it could provide some advanced features that deals with the temporal dimension of the data. For instance, excluding all the data outside a given temporal window [24]. Such extensions could allow the modelling of spatiotemporal events or flows and dynamically detect "hot spots" [32] in the stream.

But, if Spark can technically handle streaming, its implementation would raise several conceptual and technical issues. First, it would be necessary to define a spatial indexation method that support fluctuating data. Second, what would be the visual variables to use for this type of data in order to represent their dynamic structure? Those defined by Bertin in 1967 [5] and widely used since are inappropriate because of their strict limitation to a static spatiotemporal context. More recent works have tried to add visual variables to Bertin's models in order to represent motion [17, 31], but their application in a context of Big data remains unaddressed. Furthermore, once these conceptual issues are solved, the definition of a system that is effectively able to represent and manage streamed data remains to be done. This could not be a simple add-on to the classic geographic information systems (GIS): they are designed to be efficient for classical data only and are not able to deal with the huge amount of data and velocity that Big data implies. How then is it possible to manage and to represent fluctuating Big data in an efficient way, without losing the horizontal scalability offered by Hadoop? This rich problematic requires the definition of a new type of GIS. This will be the bottom line of our future research works.

Acknowledgements. We acknowledge the support of the Natural Sciences and Engineering Council of Cana-da (NSERC), funding reference number 327533. We also thank Université Laval and especially the Center for Research in Geomatics (CRG) and the Faculty of Forestry, Geography and Geomatics for their support and their funding. Thanks to, Eveline Bernier, Cecilia Inverardi and Pierrot Seban for their thorough proof reading in the writing of this paper.

References

1. Aji, A., et al.: Hadoop GIS: a high-performance spatial data warehousing system over MapReduce. Proc. VLDB Endow. **6**(11), 1009–1020 (2013)
2. Akter, S., Wamba, S.F.: Big data analytics in E-commerce: a systematic review and agenda for future research. Electron. Markets **26**(2), 173–194 (2016)
3. Armbrust, M., et al.: Spark SQL: relational data processing in spark. In: 2015 ACM SIGMOD International Conference on Management of Data, pp. 1383–1394 (2015)
4. Badard, T.: Mettre le Big Data sur la carte: défis et avenues relatifs à l'exploitation de la localisation. In: Colloque ITIS - Big Data et Open Data au cœur de la ville intelligente, Québec (2014)
5. Bertin, J.: Semiologie Graphique: Les Diagrammes. Les Reseaux, Les Cartes (1967)
6. Bray, T.: The Javascript object notation (JSON) data interchange format, RFC 7158 (2014)
7. Commonwealth Computer Research: Apache Spark Analysis (2017). http://www.geomesa. org/documentation/tutorials/spark.html
8. Cortés, R., et al.: A scalable architecture for spatio-temporal range queries over big location data. In: IEEE 14th International Symposium on Network Computing and Applications, pp. 159–166 (2015)
9. Davis, M., Aquino, J.: JTS topology suite technical specifications (2003)
10. Dean, J., Ghemawat, S.: MapReduce: simplified data processing on large clusters. Commun. ACM **51**(1), 107–113 (2008)
11. Eldawy, M., Mokbel, F.: A demonstration of SpatialHadoop: an efficient MapReduce framework for spatial data. In: VLDB Endowment (2013)
12. Eldawy, M., Mokbel, F.: Pigeon: a spatial MapReduce language. In: 30th International Conference on IEEE Data Engineering, pp. 1242–1245 (2014)
13. Eldawy, M., Mokbel, F.: The era of big spatial data: a survey. Inf. Media Technol. **10**(2), 305–316 (2015)
14. Engélinus, J., Badard, T.: Towards a real-time thematic mapping system for streaming big data. GIScience, Montreal (2016)

15. Engélinus, J., Badard, T.: Elcano: a geospatial Big data processing system based on SparkSQL. GISTAM, Funchal (2017)
16. Evans, M.R., et al.: Spatial big data. In: Big Data: Techniques and Technologies in Geoinformatics, p. 149 (2014)
17. Fabrikant, S.I., Goldsberry, K.: Thematic relevance and perceptual salience of dynamic geovisualization displays. In: 22th ICA/ACI International Cartographic Conference, Coruna (2005)
18. Fellah, A.: Une Approche Scalable pour le Traitement de Grande Quantité de Données (Doctoral dissertation) (2016)
19. Fermigier, S.: Big data et open source: une convergence inévitable? (2011). http://projet-plume.org
20. Finkel, R.A., Bentley, J.L.: Quad trees a data structure for retrieval on composite keys. Acta Informatica 4(1), 1–9 (1974)
21. Franklin, C., Hane, P.: An introduction to geographic information systems: linking maps to databases [and] maps for the rest of us: affordable and fun. Database 15(2), 12–15 (1992)
22. Gamma, E., et al.: Design Patterns: Elements of Reusable Object-Oriented Software (1994)
23. Gantz, J., Reinsel, D.: The digital universe in 2020: big data, bigger digital shadows, and biggest growth in the far east. In: IDC iView: IDC Analyze the Future 2007, pp. 1–16 (2012)
24. Golab, L.: Sliding window query processing over data streams. Doctorate thesis. University of Waterloo (2006)
25. Gu, L., Li, H.: Memory or time: performance evaluation for iterative operation on Hadoop and Spark. In: High Performance Computing and Communications & IEEE 10th International Conference on Embedded and Ubiquitous Computing, pp. 721–727 (2013)
26. Guttman, A.: R-trees: a dynamic index structure for spatial searching. ACM SIGMOD Rec. 14(2), 47 (1984)
27. Hugues, J.N., et al.: GeoMesa: a distributed architecture for spatio-temporal fusion. In: SPIE Defense+Security, p. 94730F. International Society for Optics and Photonics (2015)
28. ISO 19125-1: Geographic information-Simple feature access – Part 1: Common architecture. ISO/TC 211, 42 p. (2004). https://www.iso.org/standard/40114.html
29. ISO 19125-2: Geographic information-Simple feature access - Part 2: SQL option. ISO/TC 211, 61 p. (2004). https://www.iso.org/standard/40115.html
30. Knoll, A., et al.: Interactive isosurface ray tracing of large octree volumes. In: IEEE Symposium on Interactive Ray Tracing, pp. 115–124 (2006)
31. MacEachren, M.: An evolving cognitive-semiotic approach to geographic visualization and knowledge construction. Inf. Des. J. 10(1), 26–36 (2001)
32. Maciejewsky, R., et al.: A visual analytics approach to understanding spatiotemporal hotspots. IEEE Trans. Visual Comput. Graph. 16(2), 205–220 (2010)
33. Manyika, J., et al.: Big data: the next frontier for innovation, competition, and productivity. The McKinsey Global Institute (2011)
34. McAfee, A., et al.: Big data: the management revolution. Harvard Bus. Rev. 90(10), 60–68 (2012)
35. Mericksay, B., Roche, S.: Cartographie numérique en ligne nouvelle génération: impacts de la néogéographie et de l'information géographique volontaire sur la gestion urbaine participative. In: Nouvelles cartographie, nouvelles villes, HyperUrbain (2010)
36. Niemeyer, G.: Geohash (2008). http://geohash.org
37. Nievergelt, J., et al.: The grid file: an adaptable, symmetric multikey file structure. ACM Trans. Database Syst. (TODS) 9(1), 38–71 (1984)
38. Obe, R.O., Hsu, L.S.: PostGIS in Action. Manning Publications Co., Greenwich (2015)
39. Ram, S.: Magellan: geospatial analytics on spark (2015). http://hortonworks.com/blog/magellan-geospatial-analytics-in-spark/

40. Sriharasha, R.: Magellan's Github - issue 30 (2016). https://github.com/harsha2010/magellan/issues
41. Van, L.H., Takasu, A.: An efficient distributed index for geospatial databases. In: Chen, Q., Hameurlain, A., Toumani, F., Wagner, R., Decker, H. (eds.) DEXA 2015. LNCS, vol. 9261, pp. 28–42. Springer, Cham (2015). https://doi.org/10.1007/978-3-319-22849-5_3
42. Vlissides, J., et al.: Design Patterns: Elements of Reusable Object-Oriented Software, p. 11. Addison-Wesley, Reading (1995). 49.120
43. Vorha, D.: Apache parquet. In: Practical Hadoop Ecosystem, pp. 325–335. Apress, Berkeley (2016)
44. White, T.: Hadoop: The Definitive Guide. O'Reilly Media, Inc., Sebastopol (2012)
45. Xie, D., et al.: Simba: efficient in-memory spatial analytics (2016). https://www.cs.utah.edu/~lifeifei/papers/simba.pdf
46. You, S., et al.: Large-scale spatial join query processing in cloud. In: 31st IEEE International Conference on Data Engineering Workshops (ICDEW), pp. 34–41 (2015)
47. Yu, J.: GeoSpark's Github - issue 33 (2017). https://github.com/DataSystemsLab/GeoSpark/issues
48. Yu, J., et al.: Geospark: a cluster computing framework for processing large-scale spatial data. In: Proceedings of the 23rd SIGSPATIAL International Conference on Advances in Geographic Information Systems, p. 70 (2015)
49. Zaharia, M., et al.: Spark: cluster computing with working sets. In: 2nd USENIX Conference on Hot Topics in Cloud Computing, vol. 10, p. 10 (2010)

Author Index

Printed in the United States
By Bookmasters